Praise for *Mulberries in the Rain*

This book takes the noble craft of permaculture to a place I have yet to experience. Through the potent medicine of story, Ryan and Trevor have honored the indigeneity of this craft: that of relations. This is a brilliant text of clear, concise instructions for permaculture design, yet the stories chosen to narrate technical concepts as well as plant profiles are as rich as the recipes they gift us with. Facts are fleeting, stories live forever in the heart. Gratitude to *Mulberries in the Rain* for holding true to our most sacred form of learning.

—Kat Maier, Sacred Plant Traditions, and author,
Energetic Herbalism: A Guide to Sacred Plant Traditions

Whether you are a seasoned permaculturist, an aspiring homesteader, or simply someone seeking a deeper connection to the natural world, this book is an inspiring and practical resource. It reminds us that growing plants is about more than food or aesthetics—it's about cultivating friendships, nurturing the earth, and finding belonging in an interconnected web of life.

—Michael Judd, author, *Edible Landscaping with a Permaculture Twist*
and *For the Love of Paw Paws*

Mulberries in the Rain follows a verdant, plant-loving pathway through the human and beyond-human aspects of permaculture, weaving the personal in with the practical. Trevor and Ryan give an honest and inspiring account of their permaculture journeys with the hope that we will become more informed and inspired Earth-carers, growers, and ecological community connectors, attentive to the needs and reciprocities our own particular contexts.

—Charlie Mgee, Formidable Vegetable

This book will help you think about the possibilities for the plants in your garden. What will your stories be?

—Cindy Conner, author, *Grow a Sustainable Diet*, *Seed Libraries*
and *Homegrown Flax and Cotton*

Mulberries in the Rain is more than your typical permaculture book. This is a well-researched book—part personal narrative, part philosophy, part aggregation of the most relevant permaculture plant knowledge and techniques of the 21st century. All which blends into a highly useful, practical, and cutting-edge treatise that is fun, overwhelmingly personal, easy to understand, illuminating, and sure to expand your knowledge of how to manage plants, land, and resources—including social resources—ecologically and regeneratively.

—Blake Cothron, author, *The Berry Grower* and *Pawpaws*, and owner, Peaceful Heritage Nursery LLC

It's been my pleasure to know these fine young men and see them grow not only plants but families. They weave knowledge with experience and observation in a fresh and entertaining style, stimulating the reader towards the foundations of a society that embraces the natural world.

—Michael McConkey, owner, Edible Landscaping LLC

In place of a diversity of force—a globalism of ideas, a life without local culture or lineage—*Mulberries in the Rain* is a gentle wafting of spring air. An intriguing and conversational book beckoning forward a new position on an old trace, one gilded by the art of relationship and the heart of the conversationalist—seeking animacy, asking questions, and telling stories, without leaving the "how to" at the door.

—D. Firth Griffith, Sacred Ecologist, author, *Kincentric Rewilding* and *The Rimwalker Series*, host/writer, Unshod

A delightfully well organized and accessible book about two people's journey on the permaculture path. An easy read with a lot of personal stories, tips, ideas, recipes, and plant information on some of their favorite plants. It is particularly valuable for people new to permaculture and who are considering making a life/career commitment. The reading experience felt like sitting with a couple of good friends sharing their stories, knowledge, and experience.

— Penny Livingston Msc, co-founder, Regenerative Design Institute

mulberries in the rain

GROWING PERMACULTURE PLANTS
FOR FOOD AND FRIENDSHIP

RYAN BLOSSER & TREVOR PIERSOL

Copyright © 2025 by Ryan Blosser and Trevor Piersol.
All rights reserved.

Cover design by Diane McIntosh.
Cover image: Diane McIntosh, background ©iStock

Illustrations by Katie Garrett.
Recipe graphic: © AdobeStock_92101856

Printed in Canada. First printing May 2025.

Inquiries regarding requests to reprint all or part of *Mulberries in the Rain* should be addressed to New Society Publishers at the address below. To order directly from the publishers, please call 250-247-9737 or order online at www.newsociety.com.

Any other inquiries can be directed by mail to:
New Society Publishers
P.O. Box 189, Gabriola Island, BC V0R 1X0, Canada
(250) 247-9737

New Society Publishers is EU Compliant. See newsociety.com for more information.

LIBRARY AND ARCHIVES CANADA CATALOGUING IN PUBLICATION

Title: Mulberries in the rain : growing permaculture plants for food and friendship / Ryan Blosser & Trevor Piersol.
Names: Blosser, Ryan, author. | Piersol, Trevor, author.
Description: Includes bibliographical references and index.
Identifiers: Canadiana (print) 2024051839X | Canadiana (ebook) 20240518403 | ISBN 9781774060032 (softcover) | ISBN 9781771423922 (PDF) | ISBN 9781550927962 (EPUB)
Subjects: LCSH: Permaculture plants. | LCSH: Permaculture. | LCSH: Edible forest gardens.
Classification: LCC QK98.4.A1 B56 2025 | DDC 581.6/3—dc23

Funded by the Government of Canada | Financé par le gouvernement du Canada | Canada

New Society Publishers' mission is to publish books that contribute in fundamental ways to building an ecologically sustainable and just society, and to do so with the least possible impact on the environment, in a manner that models this vision.

Contents

Introduction	1
What This Book Is About	1
How to Read This Book........................	2
Our Story	4
Our Story Doesn't Happen Without Shenandoah Permaculture Institute	5
Our Journey Through Permaculture	7
Not Just Another Permaculture Book ..	10
Chapter 1: The Human Sector...............	13
On Plants, People, and Belonging	13
Why the Human Sector.......................	15
Ryan's PDC Story	16
Trevor's PDC Story	17
The Evolution of the Scale of Permanence	19
Arriving at a Human-Centered Scale of Permanence	20
Human Sector Tools and Strategies	23
Bringing It All Together.......................	37
Chapter 2: Food Forests Simplified	39
Trevor's Food Forest Story...................	39
Our Approach to Food Forests.............	42
Permaculture Plant Guilds	43
Trevor's Story of Growing Wild Rose Orchard	49
Site Selection and Layout....................	52
Linear Guilds	53
Linear Guild Spacing............................	55
Expanding Island Guilds	57
What Is a Permaculture Plant?.............	58
Plant Selection	59
The Food Forest Cheat Sheet...............	61
Preparation and Planning	64
Pest and Disease Management	64
Sources for Plant Materials	66
Chapter 3: Anchor Plants.......................	67
1. Mulberry..	67
2. Willow	71
3. Che	75
4. American Persimmon.....................	80
5. Pawpaw	86
Chapter 4: Nitrogen-Fixers	91
1. Redbud	91

v

2. False Indigo, aka Indigo Bush 99
   3. Black Locust.................................. 103
   4. Clover 107
   5. Goumi 110

Chapter 5: Barrier Plants 115
   1. Rhubarb... 115
   2. Jerusalem Artichoke 120
   3. Garlic.. 125
   4. Spring Bulbs 130
   5. Elderberry.................................... 135

Chapter 6: Beneficial Attractors........... 141
   1. Tulsi.. 141
   2. Zinnia 146
   3. Cannabis....................................... 150
   4. Strawberry.................................... 157
   5. Honeyberry................................... 162

Chapter 7: Dynamic Accumulators....... 167
   1. Comfrey.. 167
   2. Beet 172
   3. Nettles 178
   4. Burdock 183
   5. Yarrow 188
   We Close with a Metaphor 191

Endnotes 193
Index 195
About the Authors 201
A Note About the Publishers 202

Introduction

Whether in the context of a dream, poetic wanderings,
or a long tale of healing to tide the circle through the night,
we have learned the magic and pattern songs of the plants
through story.

Kat Maier, *Energetic Herbalism*

What This Book Is About

THIS IS A BOOK about two friends and their intertwined journeys building deep relationships with plants. We—Ryan and Trevor, the authors and the "us" of this book—both got into Permaculture and regenerative farming because of our passion for plants. That passion, bordering at times on obsession, has been a through-line for both of us as we have built our farms and homesteads and learned from our failures and successes. Plants have been like a second family for both of us and have also brought us closer together as friends. Plants have taught us how to be better humans, how to live more in harmony with the earth, and how to be easier on ourselves. Perhaps most of all, growing and learning with plants has brought us both immense joy.

We wrote this book to share that joy and passion with the world in a way that is both entertaining and informative, with the hope of inspiring our readers to cultivate lifelong friendships with plants. One of the ways we have learned about plants over the years is through storytelling, so this book is first and foremost a collection of stories. We hope that the personal stories in this book offer a framework for the reader to develop their own deep relationships with plants through stories.

This starts by treating plants as characters in our own stories—by engaging with them as members of our own interconnected communities, just like our neighbors, family, and friends. Then we tell stories—to each other, to ourselves, to the plants themselves. We share our plant stories around the fire at night, we write them down, we pass them on to our children, we ask others about their plant stories.

As you read, we invite you to reflect on your own experiences with plants. Like searching for morels in April at the base of every wild apple or tulip poplar—once you see one, you see them everywhere. How have they changed the way you view yourself or the world? How have they given you comfort, joy, healing, or laughter? How have they brought you closer to the natural world, to yourself, or to your community?

They are, after all—and it bears repeating—characters in your story, whether you realize it or not. The true work lies in the living: living with the awareness that as you come to know the plants in your life their stories will unfold. And as you move through life with a keen eye, you will find the plants revealing themselves to you. We thank you for listening to our plant stories and look forward to hearing yours someday.

For some of us just getting started on our plant journeys, however, storytelling is not quite enough. We also need some kind of practical guide for selecting and growing these amazing plants. For this reason, we have also included how-to-grow information based on our several decades of growing plants on our own farms, including a few frameworks that we have found useful in helping us grow healthy, diverse ecosystems of plants.

How to Read This Book

In this introduction, we familiarize the reader with the concept and history of Permaculture and how plants are a core part of that discipline. Through this telling, we also introduce ourselves, the authors, and the Permaculture design and education business we co-own called Shenandoah Permaculture Institute.

Chapter 1 introduces the "Human Sector"—a framework for understanding and designing human systems that is at the heart of the work we do at Shenandoah Permaculture Institute. Even though this is a book about plants, it is also a book about the relationships between humans and plants. We believe strongly that humans are part of nature, and that any discussion of growing plants must, necessarily, involve the Human Sector. What we have found over the years is that the same patterns that are useful when working with the Human Sector are also useful when working with plants—leaning in, observing and interacting, being vulnerable, making connections, and, of course, telling stories. We share what we have learned about the Human Sector as a bridge from people to plants.

Chapter 2, "Food Forests Simplified," shares a very accessible and easy-to-follow framework for growing a diversity of plants in a Food Forest system. We did not invent Food Forests, but we have spent the last two decades learning how to grow productive and efficient orchard systems modeled after Food Forests. We believe the Food Forest growing pattern is an ideal way for plant enthusiasts to engage with plants, learn how to grow them, and start building plant stories. It can be scaled from a backyard to a large farm, and it allows plants to express themselves the way they evolved to, as part of a community. In this chapter, we simplify the Food Forest concept with practical and specific ideas on how to get started.

For the remainder of the book, we go on a plant-by-plant journey, sharing personal stories about each plant followed by detailed how-to-grow information. In a sense, this main section of the book strikes a balance between stories and practical growing information. If you just want to be entertained, feel free to enjoy the stories and skip the how-to-grow. On the other hand, the how-to-grow is always there as a reference—when you want to start to get to know a particular plant, it will give you just enough information to begin your relationship in a productive way.

The structure of the plant section of the book is based on the Permaculture concept of the *plant guild*, or an intentional grouping of plants that provide functional benefits to each other so that each plant becomes healthier than if it was alone. Guilds are about diversity. We seek to build diverse plant communities much like we seek to build diverse human communities. Diversity breeds resilience.

While the concept of guilds has not necessarily been scientifically tested, we have found it to be an incredibly useful design pattern for experimenting with and learning about plants, especially for beginners suffering from analysis paralysis. Although there are different takes on the guild breakdown, we have simplified it into five categories: the anchor plant, nitrogen fixers, barrier plants, beneficial attractors, and dynamic accumulators. We have structured the book to include five plants in each of these guild categories.

Our Story

The year was 2012. That heady year when the Mayan calendar was nearing its terminal count and all the Permabros wore man buns. Trevor had a man bun, Ryan had a more ridiculous head of white-boy dreadlocks, but it wasn't hipster hairstyles that brought us together, it was a love of plants.

Early in our relationship, we would talk for hours about herbal medicine and functional plants. Trevor has a detail-oriented, exhaustive mind and an ability to dial in on niche interests within the plant world that made these conversations fascinating. Meanwhile, Ryan has an expansive mind that travels quickly across information landscapes seeking connections. Together we would saunter through conversational geography that looked like the Appalachian Mountains we both live in the shadow of—lush, diverse, fertile, edgy.

We didn't just talk about plants; we grew them together and went foraging whenever possible. One of our favorite days out in the woods happened a short while after meeting each other. Trevor knew of a spot just outside of Charlottesville, Virginia, where pawpaws dropped to the ground untouched. The mighty pawpaw had yet to garner the attention of people outside of a few hardcore foragers in the area and the certain places where they were plentiful.

We pulled up to Monticello hoping to catch a few ripe pawpaws. Ryan had yet to taste one. And on this Sunday in late August, there was nobody walking the gravel path into the woods surrounding the grounds where Thomas Jefferson once lived. The walk started out as most walks do, in conversation. We talked about Trevor's orchard design, we talked about the upcoming course for Shenandoah Permaculture, and we talked about our families. As we approached a section of the trail that Trevor knew to be a high-percentage area for pawpaws, we stepped off the trail and into the woods. Just a short walk revealed the first fruit. And after that, we couldn't unsee all the pawpaws in the woods.

That day we filled our bags to the brim, each of us taking about 40 pounds of pawpaws home. The next two weeks, our text exchange included recipes for pawpaw bread and wonderings aloud about making a pawpaw hard cider. That day's foraging would prove to be more fertile than just its pawpaw treasure. That was the walk where this book first took shape. We assembled it in our ramble through the woods from pieces put in place years ago at the beginning of the Permaculture company Trevor and I founded along with Emilie Tweardy.

Our Story Doesn't Happen Without Shenandoah Permaculture Institute

In 2010, the Shenandoah Permaculture Institute began as an idea that sprung up from a conversation between Dr. Ted Butchart and Ryan Blosser. Their friendship spilled over from a Permaculture Design Course (PDC) that Ryan took through the New Community Project in Harrisonburg, Virginia. Ted was a guest instructor.

Throughout the course, Ryan and Ted found themselves engaged in conversations around how Permaculture can impact human health through smart design. Their dialogues continually spiraled back to the fascinating observation that Permaculture tends to be, as Ted would say, "a collection of competencies." As their friendship grew, they kept returning to the idea of strengthening the approach to the human element in Permaculture.

Fast forward two years later: Ryan was now working as Executive Director of Project GROWS, a nonprofit farm in Augusta County, Virginia, working to improve the health of children and youth through food education and distribution. He was interested in training the Project GROWS farm manager in Permaculture. As we've all heard, necessity is the mother of invention, and the need for local Permaculture training soon evolved into Ted and Ryan co-leading their first course in the spring of 2013.

From the beginning, the goal of the teaching team has been to grow more Permaculture educators, and so even in this first iteration of the course, a teaching apprenticeship was created. Enter Trevor Piersol. As he stepped into the role of apprentice, his talent, enthusiasm, and knowledge were immediately apparent. As a native Virginian with a wealth of local knowledge and hard skills, Trevor shone as a natural teacher with a sturdy presence. At the end of the course, Ryan, Ted, and Trevor decided to launch the partnership of the Shenandoah Permaculture Institute (SPI).

Finally, in the fall of 2015, Emilie Gooch Tweardy, then new to Virginia, reached out to SPI to inquire about strengthening her Permaculture training, experience, and community. As it happened, another PDC was being offered soon after her inquiry. The SPI team was so excited about Emilie's enthusiasm that they offered her the teaching apprentice role in the upcoming course. During the PDC, it became apparent that Emilie's ease of communication, honed leadership abilities, and grasp of complex animal systems added to the dynamic of the team and fit in well with the spirit of feedback and collaboration. Although she'd intended only to dabble with teaching, she immediately felt a strong connection to the work and quickly expressed her interest to continue teaching with the SPI team. Late in 2015, she was invited on as a full SPI partner.

At Shenandoah Permaculture Institute, we are a collection of competencies whose mission is to inoculate communities with the tools and strategies for health and resiliency from soil to self. The diverse and enthusiastic teaching team achieves this through teaching, writing, research, and design. We teach real-world, hands-on, practical Permaculture. As of this writing, Shenandoah Permaculture has grown to include a dynamic network of alumni, teachers, and designers. By the time this book is published, we will have taught 20 PDCs in Virginia and supported hundreds of students in their Permaculture journey. Shenandoah Permaculture is the community that the ideas in this book spring from. It's our vessel for thinking and practicing concepts and strategies designed to improve our world.

The two main voices in this book are Ryan's and Trevor's. You, reader friend, will find in the stories our musings on life, death, mental health, and other Human Sector stuff. That said, there are many other characters in these stories, from our friends at Shenandoah Permaculture to the plants themselves. It is the simple act of living with and using these plants that creates the stories, and the simple act of thinking about plants as characters in your stories that creates the relationship. Like those pawpaws we discovered at Monticello, once you start thinking in this way, you can't unsee it. Plant by plant, we continue to enrich our lives as the stories keep growing, stacking like turtles.

Our Journey Through Permaculture

Here it is again, that haunting moment when someone you know across the table asks with curiosity and maybe a hint of condescension: "So what is Permaculture anyway?"

Since this is a book about stories, here's one from Ryan.

Years ago, I had just been hired by our local community services board that was operating as a fiscal agent for a food access initiative taking place on ten acres of county land. The project would go on to become Project GROWS and continues to be a powerful and ethical example of this type of work in our region. Project GROWS in its inception was committed to improving the health of children and youth in the Staunton, Augusta County, and Waynesboro region of Virginia through food education and food production. I'm proud now of the work our team put into that project and that it's still going strong without us. It's a miracle it worked because early in those years I was green, as in inexperienced. I had no idea how to run an organization and even less of an idea how to speak in public—which it turns out is just the type of thing executive directors need to do. It seemed some news media would show up weekly to our farm with an angle on a new story. One evening, they showed up with a news camera, having heard me speak previously about Permaculture, and asked me the dreaded question: "What is Permaculture?" before shoving the microphone in my face.

I freezed, y'all. It was like Will Ferrell's character in *Talladega Nights* when he's first getting interviewed and doesn't know what to do with his hands or the microphone. I almost swallowed the microphone and worse. I shit you not, I answered the question with the following words:

"Uh, Permaculture is uh, design and stuff."

Just in case this wasn't compelling, for some reason I included a second round of "and stuff" at the end.

To my horror, this answer was aired, and I promised myself from then on to always be ready with a concise definition.

Permaculture has so many definitions. Pick any guru and you'll get a different answer. While at SPI we claim anti-guru status, we still have our own definitions.

Our full definition of Permaculture, our elevator pitch if you will, is this: Permaculture is a design process using nature as a model to develop sustainable human habitats.

In Permaculture's early years, definitions and conversations around its meaning didn't always include humans or the social experiences that impacted agriculture. What it did focus on heavily was what is now called regenerative agriculture. It seems our movement does love a buzzword—until it seemingly gets co-opted, greenwashed, and robbed of its meaning. In our definition, we are intentionally not worried about using buzzwords that may have passed on in the social media's faddish pulsing of a word. Instead, we use what we think fits, and we always include the Human Sector.

One tension point that has been appropriately difficult to resolve over the years is where Traditional Ecological Knowledge ends and Permaculture begins. Specifically, the role Permaculture and Permaculture thinkers have played in co-opting Indigenous ecological knowledge under a new name. While in this book we won't attempt to resolve this tension, we do acknowledge that it exists.

Today, the two of us live on land in the Shenandoah Valley of Virginia, part of the great Appalachian Valley that stretches from Quebec to Alabama. We live about eight miles apart—our farms have the same rich limestone soil, and both drain into the Middle River, which eventually makes its way to Chesapeake Bay.

We often wonder, what stories have been told about our little section of the Great Valley by those who came before us? If we ever get the chance to hear these stories, we would be grateful, and commit to learning more.

Like most of North America, our bioregion has been continuously inhabited by Indigenous people for at least the last 15,000 years. This is a vast, dynamic history that includes a diversity of cultures, societal structures, and lifeways. To make matters more complex, the geography of the Valley makes it a natural conduit for migration and exchange. What little information we have indicates that the Valley has been inhabited over the millennia by Eastern Siouan-speaking ancestors of the Monacan and Manahoac, speakers of an Iroquoian language (possibly associated with the Owasco culture), and Algonquin speakers of the Keyser culture.[1]

We also acknowledge that there are many existing tribes throughout the state of Virginia including the Mattaponi, Pamunkey Tribe, Chickahominy, Eastern Chickahominy, Rappahannock, Upper Mattaponi, Nansemond, Monacan, Cheroenhaka (Nottoway), Nottaway, and Patawomeck.

This is a book about deep relationships with plants, and plants are inextricably linked with the land and the people they evolved with over the millennia. The stories we share here were forged on the land, and that land has a long history that includes beauty but also loss and sorrow. The plants, too, have a history. All of them co-evolved with our human ancestors, and many of the plants in this book were perhaps passed to us by Native peoples living and working on the land we now inhabit. When we enjoy the treat of a ripe persimmon in October, we are accepting the gifts of Indigenous stewards of the land we have never met. We give credit and honor to that work.

So many of the agricultural techniques and strategies used in Permaculture design and shared in this book are inspired by traditional practices of Indigenous peoples around the world. How to appropriately credit

Indigenous peoples for their knowledge and work is a complicated question, but acknowledgment is a starting point. There is so much wisdom in Indigenous ecological knowledge, and as people who want to grow a healthier, more sustainable world, we yearn to learn from it. But we do so with humility, wary of the pitfalls of co-optation and romanticization. We are reminded of one of our favorite passages from Tyson Yunkaporta's book *Sand Talk: How Indigenous Thinking Can Save the World*. This message has helped us over the years to start to reconcile our grief over what happened and continues to happen to Native Americans with our intense desire to learn and enact ancestral wisdom in our lives.

> I have previously talked about civilized cultures losing collective memory and having to struggle for thousands of years to gain full maturity and knowledge again, unless they have assistance. But that assistance does not take the form of somebody passing on cultural content and ecological wisdom. The assistance I'm talking about comes from patterns of knowledge and ways of thinking that will help trigger the ancestral knowledge hidden inside. The assistance people need is not learning about Aboriginal Knowledge but in remembering their own.

Not Just Another Permaculture Book

Confession time—within Shenandoah Permaculture, we have for years been making fun of Permaculture books. The running joke that is not a joke is that many books that come out seem to be recycled versions of the early Permaculture material. "Oh look, another Permaculture book!" we find ourselves saying in class. It always gets a laugh. Nonetheless, we've bought and read every book and proudly display them on our bookshelves, and the truth is there are gems of new insight in almost all of them. And now here we are, writing our own book!

Permaculture is many things—a set of ethics, a design system, and a checklist for decision making. Its broad, holistic nature means that, in reality, there are endless avenues for new practitioners to explore and add their unique lenses and ideas. For us, it was a quote we heard from Patrick Whitefield that inspired us to write and made us feel like we had something to add to the literature:

"Permaculture is the art of designing beneficial relationships."

When we first heard this definition, by way of the Permaculture designer Starhawk, we were floored. Permaculture is notoriously difficult to define, and those brave enough to attempt usually end up getting lost in a word salad. This definition is so simple and, for us, gets right to the core of what Permaculture is about. After discovering this idea, we finally arrived at feeling like we had something to say and envisioned the kind of Permaculture book we wanted to write. What we landed on was Permaculture in its most poetic form. The inspirational quote became paraphrased in our courses to be even more powerful:

"Permaculture is the art of relationships."

Across the decade we've been teaching together, we find students come to our courses eager to build their own plant competencies but unsure of where to begin, and so we teach them that the core of building personal plant knowledge is to form a relationship with the plant.

"You don't really know a plant until you've had a conversation with it," we often say.

In our courses, we share stories of our own plant relationships and how they tend to be woven into the fabric of our human relationships. The positive responses we've received to those stories became the idea for this book.

This book is not just about plants. You will find some tools and strategies we have used and updated in our Permaculture courses. One thing we have worked hard at while refining our Permaculture curriculum is the Human Sector. The first part of the book, where we detail and explore Human Sector ideas and strategies, is the bridge to the second part, about plants. We build our plant community simultaneously with our human community.

CHAPTER 1

The Human Sector

On Plants, People, and Belonging

We walked into the room at the University of Richmond to a packed house. Half the people in the class were wearing masks, the other half weren't. In all, 40 people packed into a tiny classroom. It was the tenth course we had taught and our first since COVID. It was by far the largest course to date. We looked at each other with nervous glances.

Things started shifting just before COVID. When we first started teaching, our courses were small and homogenous. Our classes were filled with a certain type of left-leaning do-gooder. Mostly white, usually young, often wealthy or at least upper middle-class. By now, though, the makeup of our courses is becoming more diverse. Ideologically, we are seeing more people from the so-called right, more Christians, and a lot of veterans. Still mostly white, but less so. In addition, the courses are now multigenerational and tend to be made up of people less on the fringe and more folks feeling trapped by conventional workday life. COVID seemed to speed up this transition to more diversity in our courses, and it did something else. It made people hungry and anxious for change.

The packed room that day in Richmond had an energy to it. Nervous, thick, and full of both creative and destructive potential. We got the sense that if someone lit a match, the whole thing would blow.

A man in the front row in his 70s, with long stringy hair, faded tattoos blurred by the sun, and a cigarette pack hanging out of his pocket glanced around the room. A woman in her 30s made eye contact with me and said, "I don't know if I'm ready for this." She didn't intend to speak for the group, but it was obvious, everyone was feeling the same. Whether they admitted it to themselves or not, sharing a small space with 40 strangers after the lockdown we'd all had was weird!

We had spent the last two years in our bubbles, with intimate friends and family or, worse, alone. Half the people around us thought the world was unsafe, and the other half believed people who were scared were crazy. The nervousness had an undercurrent of distrust.

Then it hit us—the feeling, the vibe, why were we all here? We didn't have to ask. Ryan opened that course with this sentence.

"You are all here because you want to belong."

The room exhaled and for the rest of the course we got to work, learning about each other while we learned about Permaculture. As we reflect on this first course we had after COVID, it becomes apparent that belonging is what Permaculture is about. It's about belonging to our landscape as much as it is about belonging to a community. A surface glance leads us to believe that these are separate, but this changes when we look deeper.

During one of the icebreakers that day, a student asked us a question. This came on the heels of Ryan talking at length about his favorite plant, the beet.

"How do you know so much about so many plants? I don't even know where to begin!"

Early in our respective plant journeys, we would memorize plant catalogs. Any plant nerd reading this understands this comforting urge. Just after Christmas, on a snowy day, Ryan liked to open a bottle of top-shelf whiskey, get a good fire going, and spread out plant nursery and seed catalogs in front of him. That ritual always kicked off the season.

The dream, the fantasy of plants to come, fed our spring and summer work. While this certainly paints a pleasant picture, the approach felt inadequate. Neither of us enjoyed the rote memory of it, the ability to spit out "back of the baseball card" facts about plants we had no relationship with. Instead, we longed for something deeper. Through conversations about this longing, we both landed on the fact that working and living with plants delivered the thing we longed for—something more than facts—it was a relationship. A sense of comfort and ease with a plant due to having spent so much time with it.

This discovery came on the heels of shared stories, be it Trevor talking about his relationship to strawberries and how this is intimately tied to his son or Ryan's relationship with cannabis and the funny, traumatic history he has with the plant. We realized that, in both of our lives, plants were more than just a theme song; instead, we were the theme song for plants. Much like an ecosystem, this interdependent story is the story.

The idea of belonging and building plant relationships snapped into focus.

Plants, like music and food, feed our culture. They create our community; they *are* our community, and this book offers a model for how to build both plant and human communities. Story by story—or rather, plant by plant—or, to quote a student on the first day of our post-COVID course years ago: "I came for the plants, but I'm staying for the people."

Why the Human Sector

This chapter is about how to think about being a better human; the next chapter is about how to think about growing a Food Forest; put them together and you have plant stories.

But why the Human Sector?

A good friend and colleague of ours, Laura Mentore, is an anthropologist at University of Mary Washington. She took our Permaculture course and completed a teaching apprenticeship and is now a lead instructor with Shenandoah Permaculture. Last year she taught a university course that included a Permaculture-focused curriculum entitled "Anthropocene: Designs for Living in the Climate Crisis." The "Anthropocene" is just a fancy word for the geological time period where humans are having a substantial impact on our planet (and a mostly negative impact at that). We're aware that this is a controversial concept among academics. Nevertheless, we fully believe we are living in the Anthropocene, which makes the Human Sector even more important.

We define the Human Sector as human energies impacting a landscape. We noticed early in our Permaculture journey that serious consideration of Human Sectors seemed to be limited, and so we made a conscious decision to focus our work on researching and integrating Human Sector thinking into our courses. To return to the opening statement, Human Sector thinking is all about how to be a better human. We both arrived at this realization at separate times.

Ryan's PDC Story

I experienced my first Permaculture Design Course (PDC) in 2011. My wife, Joy and I were living on what would become our market farm, Dancing Star Farm, and I was working as a child and family therapist serving families throughout my region. I had just finished my graduate work in clinical mental health counseling and so was approaching the world through a family and human systems lens.

In addition, I took the course on the tail end of a failed cooperative farming experiment. On my land, a group of friends got together to ride out the Great Recession, driven by the energy and fantasy of collective farming. We went hard and fast only to crash in a pile of mental health tragedy and infighting over accountability. I learned so much from this failure but had yet to understand it through either the human systems lens or Permaculture.

I came out of that first PDC inspired and armed with the kind of robust tools for design work and, as important, the practical experience of slowly walking through the design process with a group of learners. I also had a great time and developed relationships with other students in the course that are still going strong today. I quickly understood that one of the most powerful aspects of the course was the relationship piece. For a brief time, the course felt like a community, and this experience was healthy for me on the heels of the social failure I had just participated in.

At the time, language like "social Permaculture" was being used all over social media and by budding practitioners. Meanwhile, a term started popping up that I personally didn't understand. This term was an addition to the zoning on a design site. In a design site, zone 1 is the landscape immediately around the house. The zoning moves from the house out through zone 5, which is intended to be an untouched landscape. It's a useful thinking strategy for space and placement of elements in that space. Some Permaculture thinkers added zone 0 for the house and zone -1 or zone 00 that was intended to be the inner landscape of the designer.

I understood what thinkers were trying to do; however, the framework seemed more clever than accurate. Here's why: zones are used in Permaculture as a model for how to order elements in space and proximity based on time spent/resources needed whereas sectors are about energies acting on a site. Human energies affect all zones, therefore putting social Permaculture in a zone like zone 00 or zone -1 strikes me as a thinking error. We are living, after all, in the Anthropocene, for good or for ill. The Human Sector permeates everything.

What also lingered was the feeling that something was missing. It seemed like only lip service was being given to the Human Sector or social Permaculture. During the first course I took, an instructor split everyone into groups and said, "You know, just use your people skills," when referring to a conflict that one of the groups was having. This felt inadequate among the backdrop of the articulated need for "more social Permaculture."

Fast forward to the creation of SPI; we all agreed there is an opportunity to layer in a more robust conceptualization and practical understanding of the Human Sector.

We started by adding this to the Scale of Permanence and have continued to build from there.

Trevor's PDC Story

In the PDC I took with the Blue Ridge Permaculture Network in 2012, we were lucky to have professional Permaculture designer Dave Jacke, author of *Edible Forest Gardens*, as one of our lead teachers. He taught us a version of the Scale of Permanence that expanded on P.A. Yeoman's original version to make it clearer, all encompassing, and with a much needed and missing section on aesthetics.

After my class, fueled in large part by Dave Jacke's enthusiasm for Food Forests, I became obsessed with learning about plants, particularly edible and fruiting perennials. This led me to spend two years living on a mountain in rural Virginia with a cohort of eight other young adults growing gardens and orchards, raising livestock, foraging, cooking, having the time of my life, and meeting my future wife Jenna.

I went up to that mountain on an urgent quest to catch up on the hands-on experiential learning I felt like I had missed as a child of mainstream America. I wanted to finally learn how to live as part of nature, not separate from it. I wanted to build an intimate relationship with the plants and animals and soil, beyond the classroom theory I was used to, in a way that was intimate and visceral. What I didn't know I wanted, but discovered along the way with those eight adults in my cohort, was how to live in better relationships with humans as well.

When I came down from the mountain, I spent three years developing a community educational farm, continuing to learn and grow. Again, I was driven by my love of growing plants, but what I learned was that it was not enough just to grow healthy plants—it was the relationships between all the humans involved in the project that could make or break the whole thing. I was reminded again of something Dave Jacke had said along these lines: "95% of all the failed Permaculture projects I've worked on failed because of poor design of the Human Sector."

This was the context and experience I brought with me when I first met Ryan and started teaching with him. Of course, as a former child and family therapist, Ryan had his own unique insight into the importance and challenges of the Human Sector. During our first PDC together, we brought out Dave Jacke's version of the Scale of Permanence—an incredibly useful checklist for organizing design thinking—and it suddenly hit us that it was missing a section about humans! We got together and revamped it, deciding to put the Human Sector towards the top, just below climate, because of its importance and how difficult it can be to change or influence.

To be clear, Dave Jacke and many others before us have long stressed the importance of what has been called social Permaculture or the "invisible structures" of design. Permaculture is a discipline that evolves organically as different practitioners add their unique lenses to the field. What we simply did, and continue to do in our classes, is emphasize the vital importance of Human Sector design in all things Permaculture, starting with placing it at the top of the beloved Scale of Permanence. In addition, we have over the years fleshed out each section to make the checklist even more robust and exhaustive.

The Evolution of the Scale of Permanence

The Scale of Permanence first popped up back in 1958 spilling out of the brain of P.A. Yeomans in his book *The Challenge Of Landscape: The Development and Practice of Keyline*. It was a tool, or rather, a thinking strategy to help organize an order for planning based on relative permanence in the landscape. Yeoman's order was the following:

1. Climate
2. Landscape
3. Water Supply
4. Farm Roads
5. Trees
6. Permanent Buildings
7. Subdivision Fences
8. Soil

In both of our courses, we used Dave Jacke's updated version of the Scale of Permanence, which looked like this:

1. Climate
2. Landform
3. Water
4. Access and Circulation
5. Vegetation and Wildlife
6. Microclimate
7. Buildings and Infrastructure
8. Zones of Use

9. Soil (Fertility and Management)
10. Aesthetics

This ordered way of thinking chunked together the task or checklist of both observation, analysis, and implementation. It helped us to think about the whole in actionable parts and piece them together. Below is the updated Scale of Permanence that we at Shenandoah Permaculture have landed on followed by explanations of the new additions' importance and how we use them.

Arriving at a Human-Centered Scale of Permanence

SPI Scale of Permanence Checklist

By Trevor Piersol, Ryan Blosser, Emilie Tweardy

Adapted from P.A. Yeomans and Dave Jacke

Climate
- Plant hardiness zone
- Predicted future climate change status
- Annual precipitation
- Seasonal distribution
- Latitude
- Wind directions: prevailing, seasonal variations, storm wind directions
- Average frost-free dates
- Chilling hours (important for fruit tree dormancy)
- Extreme weather potential: drought, flood, hurricane, tornado, fire

Humans / Social
- Ecology of Self
  - Intrapersonal
  - Interpersonal
  - Transpersonal
- Human in-and-out (head, hand, heart)

- 8 Forms of Capital analysis
- Project stakeholders
- Neighborhood and community
- Population: density, demographics, patterns
- Cultural activities and uses
- Current uses by neighbors and passersby
- Legal limits: property lines, conservation easements, zoning, rights-of-way, setbacks, environmental regulations (e.g., protected wetland), agricultural and forest districts
- Site history: past uses and impacts on land
- Future potential uses (e.g., local economic development plan)

Landform
- Slope (steepness, rise/run in percent)
- Topographic position (i.e., mid-slope, hill crest, valley floor, etc.)
- Bedrock geology: depth to bedrock, type of parent material, pH
- Estimated seasonal high water table depth
- Elevation
- Landslide potential

Water
- Existing sources of supply: location, quantity, quality, dependability, sustainability
- Network layout and features (spigots, pipes, filters, etc.)
- Watershed boundaries and flow patterns: concentration and dispersion areas, including roof runoff patterns, gutters, and downspouts
- Potential pollution sources: road runoff, chemical runoff from neighbors, etc.
- Flooding, ponding, and puddling areas
- Possible sources of supply: location, quantity, quality, dependability, sustainability, cost

- Location of all on-site and nearby off-site culverts, wells, water lines, sewage lines, septic systems, old wells, etc.
- Erosion: existing and potential areas

Access/Circulation
- Activity nodes, paths of use, storage areas
- Pedestrian, cart and vehicle access points, current and potential patterns
- Materials flows: mulch, compost, produce, firewood, laundry, etc.

Vegetation and Wildlife
- Existing plant/fungi species: locations, sizes, quantities, patterns, uses, whether poisonous, invasiveness, weediness, what they indicate about site conditions, etc.
- Existing animal species: diversity, population size, pests
- Keystone species: e.g., old-growth trees, mycorrhizal fungi, large mammals, predators, etc.
- Stages of ecological succession
- Ecosystem architecture: layers, patterns, diversity, light/shade, character, quality
- Habitat types, food/water/shelter availability

Microclimate
- Define various microclimate spaces
- Slope aspects (direction slopes face relative to sun)
- Sun/shade patterns
- Cold air drainage and frost pockets
- Soil moisture patterns
- Precipitation patterns
- Local wind patterns

Buildings and Infrastructure
- Building size, shape, locations of doors and windows, exits, and possible functions

- Roads/pavement
- Power lines (above and below ground) and electric outlets
- Outdoor water faucets, hydrants, downspouts, septic fields, wells
- Location of underground pipes and utilities: water and sewer, gas, fiber optics, drain tiles
- Fences and gates

Soil Fertility and Management
- Soil types: texture, structure, soil profile, drainage
- Topsoil fertility: nutrients, pH, % organic matter, soil food web
- Hardpans or impermeable layers of soil
- Soil toxins: lead, mercury, cadmium, asbestos, etc.
- Management history

Aesthetics/Experience of Place
- Defined spaces (walls, ceilings, floors), qualities, feelings, functions, features
- Arrival and entry experience: sequencing, spaces, eye movements, feelings
- Specimen trees/landscaping (e.g., owner's favorite rose bush)

Human Sector Tools and Strategies

The Ecology of Self: Towards a Conceptual Model

The Ecology of Self is a model developed out of needing to consider the inner landscape of the human energies impacting the land as much as we need to understand the outer landscape. Just like we do with the land, we want to think through and map out our observations of our inner landscapes. Once we have them mapped out, we can analyze them and begin to understand the opportunities and barriers that exist for our projects and our site to meet the goals we have set. The Ecology of Self is simple.

1. **Intrapersonal:** The way the designer relates to self.
2. **Interpersonal:** The way the designer relates to other people
3. **Transpersonal** (ecosophy): The way the designer relates to the experience of being part of something larger than themselves.

The above model is a thinking strategy. Much like the Scale of Permanence, it lists a way to systematically understand the humans in the system and analyze how they may impact the design. For each part of the Ecology of Self, information can and should be gathered in order to create a more complete perspective on the project and site.

We can remember the feeling of impatience that almost always leads to wasted resources, for example, by planting a tree in one spot only to move it the next year. For both of us, when we decided to farm, we wanted to farm immediately. This desire for action is normal, but it is flawed. One balancing salve to this restless energy to act is the Scale of Permanence. The Ecology of Self as a conceptual model fits right into this work.

Our action in this early work is observation and analysis. The characteristic needed is only curiosity about your site and yourself. Warning: Some of this work may take a long time.

Intrapersonal. One may start the inquiry by asking what your self-talk is like. A great way to understand self-talk is to reflect on the things you say to yourself in your own head in those quiet moments of struggle. Are you an inner critic or an inner coach?

For example, it could be: "I hate myself, I'm so stupid," vs. "I'm having a hard time right now, but I've been here before and things will get better." How does this impact resilience and grit? What is your track record with accomplishing hard things? How confident do you feel at any given time? What are your triggers and how do you respond to them?

Interpersonal. Start with the general guiding question of what your relationship patterns look like. How has your history with setting boundaries been? How has your history with establishing relationships been? Do you have a love language? What do others do that makes you feel safe? What behavior do you find yourself engaging in, in order to get your needs met? What are your needs? What do others do that makes you feel threatened? What role do you often fill when engaged in teamwork?

Transpersonal. Start with any wisdom tradition you may use to orient yourself to the world. For many this will be Christianity, for others Islam, or Buddhism. Bill Mollison, who along with his graduate student David Holmgren is credited with the development of Permaculture, has a funny line about gatekeeping Permaculture from becoming "spinning woo-woos." The dude had his opinions about the spiritual New Age and what it might offer Permaculture. We have taken a different approach to this and have concluded that it is entirely appropriate to bring spirituality into Permaculture through this model.

On this, we have been inspired by the thinking of deep ecologist Arne Naess and his concept of ecosophy. We feel slightly self-conscious to be pulling another portmanteau into a book about Permaculture—yes, we should make fun of the inherent laziness that can be found in the making of a new word by combining two words. But still ...

Ecosophy is the combination of the words "ecology" and "philosophy." Ecosophy describes the act of living in equilibrium, harmoniously with our environment. It lends itself to a thinking that centers the connection between the self and the environment. Simply put, the act of caring for the Earth becomes, or rather is, an act of caring for the self. The path to this holistic thinking starts with the idea that there is something bigger than us. Our wisdom traditions and spiritual practices are a pathway towards a holistic viewpoint that allows humans to see Earth care as a radical act of self-love.

It's a tricky tightrope walk, this self-centered approach, and largely counterintuitive. Put another way, by expanding the circle of what and how we identify with the world, protecting the Earth and its ecosystems becomes self-care. This expansion requires deep listening and an increased awareness. The need to expand our awareness helped us choose the tools we are sharing in this section, starting with the LUV triangle.

We recommend keeping a journal as you start this process. After all, writing is thinking—writing leads to storytelling, and storytelling is, well, everything. One of the most powerful uses of narratives regarding the Human Sector is realizing after analysis that you can change your own story. Permaculture is about transformation so why not extend this to the self?

In addition to conceptual models that support our thinking strategy, we have gathered into the Human Sector tools and techniques for observing and interacting. The following are just a few of many that we rely on and teach in our courses.

Tool # 1: The LUV Triangle

The LUV triangle is a strategy and skill set Ryan learned from his mentors Renee Staton and Ed McKee in clinical mental health counseling at James Madison University.[1] The strategy is a foundation for counselor training and one that we started to use in our courses to support students in their thinking about how they interact with each other in class, in their team projects, and with potential stakeholders and clients when designing sites.

We teach this in our sessions to provide skill exposure and practice with tools that enhance community and Human Sector design.

LUV is an acronym for Listen Understand Validate. At the end of the day, this technique is best understood as structured curiosity. Be curious—that's the point. This curiosity is what helps the speaker or sharer to feel attended to and acknowledged.

L—Listen

The first piece of LUV is listen. On the most simple level, it is silence. As the listener seeking to understand someone, you start by listening with your mouth shut. Understanding someone else has nothing to do with your ideas. Be patient while listening, often it takes a while for someone's ideas to be clarified. Wait for it, ask questions, if you don't understand, say so in a way that leads them to believe you are curious. Ask open-ended questions like "What was that like?" and make eye contact. If eye contact is difficult for you, focus on the bridge of their nose or the top of one of their ears. Don't fidget.

When listening, be aware of personal space and body language. Ryan is a large man, he rarely stands close enough to touch someone, and he never squares his shoulders; instead, he stands or sits at an angle that softens the interaction considerably. In addition, if you are someone who crosses your arms or legs for comfort, don't. Crossed legs can indicate that a listener is not in a place to listen.

For many of us, crossing our arms is a form of self-soothing. If you need to do this, try crossing your toes. The listener won't notice this.

Avoid being a bobblehead. This is hard. For many of us, once that head starts nodding, it's like it's on a string. Instead, try a technique called the Lassie twist, where you tilt your head slightly to the side like the character from the 1950s TV show *Lassie*.

U—*Understand*

With LUV we are seeking to understand. Our own point of view is set aside, and we are curious. Try to repeat what you think the person is trying to say. Start with phrases like "I think you're saying that …" or "Correct me if I'm wrong please, what I hear is…"

And when in doubt, there is a magic phrase that keeps people talking. It is simply, "Tell me more."

The important piece to remember about this aspect of LUV is that not understanding is not a deal breaker. Be open to feedback when the person sharing says, "No, I'm saying this." Not understanding is an important signpost on the road to understanding.

To continue to understand, do the following: repeat what the listener says and use phrases that are similar to the listener.

Don't forget the metaphor.

A metaphor can show that you, the listener, understand the speaker in a powerful way. As a listener, the conversation is not about you, and, more importantly, you aren't trying to solve a problem or offer solutions. Instead the listener wants to understand and then communicate that understanding. Metaphors are key to this. Phrases like: "It sounds like you're a butterfly trying to break free from your cocoon," or "Wow, it's like your energy is like a river being dammed up," or another, "Wow, you've really been running a marathon with this work lately."

V—Validate

This kind of validation is what we are getting at with this work. It's important to remember that agreeing with someone's ideas and validating their experience are two different things. We see a lot of ideological diversity in our courses, and our goal as instructors is to keep the conversation going.

This is important: Don't communicate skepticism or doubt. Don't debate.

When we practice this work in our courses, sometimes students get worried about working with someone who doesn't have their same ideological beliefs. Remember that the fact that someone is struggling or hurting or happy about something is not the content of the idea—it is the psychosocial experience of the speaker. Validate that shit. You can talk about ideas later. This can be done using metaphors or by simply recognizing the speaker's feelings and repeating what you heard with phrases like the following.

"I can tell that is hard for you." Or, "It sounds like you're pissed about that." Or, "You're really jumping for joy today!"

Tool #2: 8 Forms of Capital

At SPI we teach the 8 Forms of Capital. It's a brilliant conceptual framework for thinking about economics. I was first introduced to this concept through the writings of Ethan Rowland and Gregory Landua.

In thinking about economics, and by pulling ideas from the social sciences, they expanded the understanding of capital beyond finance into other realms. The eight forms they landed on are the following:

Financial
Social
Living
Intellectual

Experiential
Cultural
Spiritual
Material

It never fails. Whenever we teach this, folks always feel drawn towards value judgments about which form of capital is superior to the

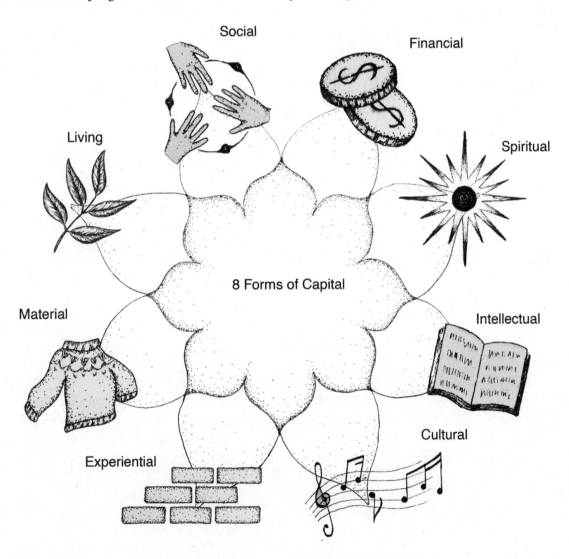

other. Often financial capital gets put at the bottom and social capital gets placed at the top of the pile. This, I believe, misses the point.

Don't get me wrong, I get it. Social capital is the feel-good one, and some Permaculture practitioners in all of our upper-middle-class wisdom like to demonize and discount the importance of financial capital. In Rowland's conceptual framework, he does a nice job of exploring how each form of capital has the potential for deficits and credit and that this creates opportunities. The brilliance of the 8 Forms of Capital is that it is a tool for analysis and design. Much like the way we use the Scale of Permanence, we now have a tool to analyze our economic landscape and use that tool to discover how to intervene, i.e., how to live a more balanced life. For many, this might mean how to live a life without such a high need to earn.

Financial
Financial capital is what most people understand as conventional capital. It's money. Included in this analysis would be debt—credit card, student loan, mortgage, etc.—and cash flow. Conventionally speaking, when we think about economics, we think about financial capital.

Social
Social capital is sometimes misunderstood as generosity. As we talk about social capital in this model, it refers more to political capital. It's the ability of someone to influence a decision or to make something happen in a community.

Living
Living capital includes the living/organic systems you have access to. Fruit trees, gardens, land, natural resources, etc. We include national forest or public lands in this.

Intellectual
Intellectual capital is on a spectrum from problem-solving ability to education and training. We also include books or a library in intellectual capital.

Experiential
Experiential capital is the experience a person brings to a community. Be it raising kids, farming, building, or teaching. It's the thing we have done for most of our lives.

Cultural
Cultural capital is the collected efforts of a community. It's art, it's songs, coming together in celebration. Cultural capital is the recipes that are passed down through the community. Cultural capital cannot be gathered or spent by an individual.

Spiritual
Spiritual capital is the thing that gets a person out of bed—the well they draw from to get them through the day. Simply understood, it's a person's purpose. As a teaching team, we always get stuck on this part, none of us being very religious. It was a student named Kelsey who helped us put words to this.

Material
Material capital is the material world we have access to. Shoes, cars, housing, clothes. The things that make up our lives.

By analyzing each of these categories, a person can start to understand where they may be trading one form of capital for another that isn't bringing their life into balance. For example, someone may go into debt for a new pair of Nikes—despite my environmental positioning, I'm a sucker for a new pair of Jordans. Go ahead, judge me.

When this happens, material capital is high and maybe this inches up social capital in some circles, but it also increases a person's need to earn in a job they may not like that decreases spiritual capital, never mind the impact it may have on living capital in other countries (i.e., sweatshop labor). This is a system out of balance.

There are so many ways to apply the 8 Forms of Capital to a system after an analysis. Often, we are doing this without thinking about it.

At Shenandoah Permaculture, thanks to our friend and co-owner Emilie Tweardy, we are big on potluck culture. In the class, the invitation starts small with the first weekend where we invite students to bring in snacks, pickles, or any other homemade/home-canned foods to share on the snack table. People are so proud of their jams, jellies, and ferments, and this is a great way to showcase them and invite the class to start making connections. In every course, and occasionally in the summer, we hold a giant potluck and seed swap.

At first glance, I love this for what it is—a big-ass party. On second glance, it is so much more.

At this giant party, students meet alumni, friends are meeting friends. Plants are exchanged, recipes traded, food shared, seeds swapped, and most of all, stories are told. We recall one particularly magical potluck where our friend and colleague—one of the best farmers we know—Betsy Trice, who farms at Peacemeal Farm with her husband, Chris, showed up with a groundhog, aka, whistle-pig, cooked down in wine sauce. This whistle-pig au vin tasted incredible, and everyone got a chance to wash it down with Betsy's homemade elderberry wine. Later that night, stories were swapped around the fire about the source of the groundhog. Turns out Betsy teaches a sustainable farming course pathway at Reynolds Community College, and the college farm needed a groundhog dispatched. About as resourceful as anybody we've ever known, Betsy of course didn't want the dead groundhog to go to waste. So, she skinned and butchered the animal. A hop, skip, and a jump later, we've got groundhog au vin.

If we were to trace the web of interactions happening across the 8 Forms of Capital in this anecdote, the diagram would be rich in intersecting pathways. Aside from modeling and living the very real but played-out Bill Mollison quote, "the problem is the solution," Betsy's act and all that went into it—trapping, butchering, recipe creation (she even made a hat with the skin for her husband)—and the stories that were told from this added wealth to the community at that potluck.

That act was a gift of intellectual capital, living capital, spiritual capital, social capital, and experiential capital but very little financial capital. Betsy used her wealth in other areas to add wealth to the community. In exchange, she used very little financial capital to achieve this. Others may have used financial capital, in the form of buying food, to contribute to the potluck, which is a perfectly acceptable though admittedly less charming approach. In a potluck experience like this, stories, the money spent, the food made, the recipes shared all contribute to balancing a community experience where everyone can get their needs met without needing an abundance of one single type of capital.

In thinking through the 8 Forms of Capital, we have another example: my transition from the role of Executive Director at Project GROWS to full-time farming.

The decision to leave Project GROWS didn't exactly come all at once. I had been thinking about the next thing as early as my second year growing the project. This thinking included the knowledge that Jenna Clarke, our director of operations, was much better than me at leading nonprofits—she knew more, worked harder, and loved people more. It would be the perfect role for her.

Meanwhile, the nudge that ended up being the thing that moved me in the direction of starting my own market farm project was the money and the struggle. I loved Project GROWS and sunk every ounce of effort I had into growing the nonprofit. This included making sure staff would get paid on time even if that meant I did not get paid. As the leader of Project GROWS, I missed an incredible number of paychecks, and we (my family and I) were broke.

If I was going to be broke, I figured I might as well be running my own farm. So, I moved on and launched Dancing Star Farm. Jenna took over the reins of Project GROWS and turned it into an incredible, functional organization.

Being a nonprofit leader in our area didn't make me any money. I think my highest earning year was in the vicinity of $20,000. But the opportunity increased my general wealth when considering other types of capital.

The work increased my profile considerably in the region. As executive director, I was on the news constantly speaking about food insecurity or answering questions about farming. In addition to this, I was able to interact with our community constantly. This created several connections

Whistle-pig au Vin

Ingredients:

1 (5 to 6 pound) groundhog
2 tablespoons extra-virgin olive oil
1¾ cups chicken broth
2 medium onions
3 cloves garlic
1 teaspoon fresh thyme
¾ stick unsalted butter
1 bottle Bordeaux
⅓ cup Dijon mustard
1 tablespoon sea salt and pepper

Directions:

Cut groundhog into pieces, rince, remove any fat, and cut out the glands underneath the front legs and armpits, then pat the meat dry. Season with sea salt and pepper. The groundhog will be unpleasant if this step is skipped—for real; do this or else.

Heat the oil in a cast iron skillet, then brown the meat.

Move the meat to a medium heavy pot.

Add broth to the pot.

Pour off any fat in the skillet, then add onions, garlic, thyme, and 3 tablespoons butter.

Add the wine and bring to a boil.

Pour the mixture over the whistle-pig. Cover the pot and bring to a simmer. Braise whistle-pig until tender, usually 2½ hours.

Remove whistle-pig from pot

Bring liquid in the pot to a boil and reduce it by half.

Whisk in the mustard. Remove from heat.

Stir in the rest of the butter.

Pour sauce over whistle-pig.

for me. In terms of capital, my social capital was extremely high at the end of my tenure at Project GROWS. In fact, I am still reaping the benefit of this almost a decade later.

In addition, I was given experiential and intellectual capital in this role. Project GROWS allowed me to learn the craft of farming at scale along with sending me to trainings and conferences where more connections were made.

During the transition from nonprofit leader to market farmer, I can remember being worried about not having enough customers. That's when I realized the value of social capital. The first paying customers to my farm were made up almost entirely from people in the Project GROWS community. They knew what I gave to the organization, and they showed up with support when I needed them the most. In this scenario, the thing I needed the most—financial capital—but had the least, showed up in my life because of my wealth in another category. By understanding how this works, the 8 Forms of Capital can be used to design our life around our goals, reducing the need to earn.

In the same way that this concept or thinking strategy can be used to analyze one's economic experience, it can also be used to build a better-balanced life.

Decisions have consequences. And this tool tells a more holistic story of the consequences behind the decisions.

Tool #3: Human In-and-Out

For many of our readers who are already familiar with Permaculture, we are confident you have come across the famous chicken in-and-out exercise. It's a classic Permaculture thinking strategy where a chicken is drawn on a white board and inputs are brainstormed for what the chicken needs. It may look like but is not limited to something along the following lines: food, exercise, other chickens, sunshine, roost, dust, grit, etc.

Next a list is made for the chicken's outputs. This list may look like this: meat, eggs, calcium in the form of feathers, manure, scratching, entertainment.

This analysis can then be done on other elements in a design before finding where the two can connect. For example, one might then do an input-and-output exercise with a greenhouse. Through this work, very quickly we'd discover that a greenhouse needs heat, provides shelter, and then it snaps into focus. Put the chickens in the greenhouse.

We love this kind of work; for us it represents a path for how to arrive at elegant solutions that often hit our personal favorite Permaculture principle: integrate don't segregate, aka stacking functions. The key to stacking functions well is making sure that every element added to a system has two or more functions and is supported by two or more elements. This underscores a core Permaculture strategy of making sure to always connect—also a core imperative when thinking about relationships.

This strategy has several applications in the Human Sector. We can run an inputs-and-yields exercise on individuals in a system by using the 8 Forms of Capital that we previously discussed. This is how to make connections about needs and supports in a human community. Another form of the Human in-and-out exercise that takes place within a broader community we call Head, Heart, Hands.

Head, Heart, Hands
On the first day of every course, we assign the Head, Heart, Hands homework. The challenge is to spend the evening documenting five things that the student is gifted in for each category. For example, Trevor's first pass at this might look like this:

> Head: organized, ordered thinking
> Heart: loyalty
> Hands: planting/pruning trees.

He would then repeat this list five times, assigning a new characteristic or skill to each category.

For Ryan it might be:

> Head: poetry writing
> Heart: intuitive listening
> Hands: athletic; good at ball sports/water sports.

Conversely, we ask each student to document five needs or "debts" they might have in each category. Trevor's might look like this:

Head: jokes to cut the seriousness
Heart: listening
Hands: earth moving/digging.

When each student arrives in class the next morning, we open by taking an inventory. Once we take the inventory, then we start making connections. Almost always the group of strangers/emerging learning community comes to a very real understanding that everything we all need is already in the room. The played-out quote, "We are the ones we've been waiting for," snaps into focus and is refreshed when participating in this exercise in a large group.

Bringing It All Together

We've shared a lot. The Human Sector is huge and still largely unexplored in a formal way through the Permaculture lens. What we have chosen to share is just the tip of the iceberg. We anticipate this changing and growing as the conversation around Permaculture snowballs to include people new to the concept and those already thinking and writing about it. In this section, we have included a model, the Ecology of Self, along with a select few of the many tools we use in our courses to explore and build skill around the Human Sector. In addition, we have shared where we have chosen to add the Human Sector and beef it up in the Scale of Permanence. This work is incomplete at best, and we look forward to expanding it in future writings. For now, let the following truism suffice: when we don't pay enough attention to the Human Sector, our designs fail. The work of partnering with the landscape to build a better, more ethical habitat is also the work of being a better human.

CHAPTER 2

Food Forests Simplified

Trevor's Food Forest Story

WHEN I WAS IN COLLEGE, I thought Permaculture was a weird, unscientific movement bordering on a cult. I had just gotten passionate about organic farming after volunteering at a small campus garden started by some of my friends. The next summer, I grew my first Cherokee Purple tomato in the red clay soil behind my apartment, and I was instantly hooked. I had been unsatisfied with the abstract world of academic theory for years, and this was finally something I thought could make a tangible difference in the world. Organic gardening, I believed, was a way to grow nutritious food while improving the health of the soil and ecosystem—plus it was really fun!

But Permaculture? This was a bridge too far for me. I had one friend who had recently became obsessed with Permaculture and was trying to start a local group on campus. But he was one of the more politically left-wing guys in our group, and so I wrote Permaculture off as too radical for my tastes. After all, I was still a recovering left-brain academic type, and I was interested in "science," not what I saw as hippie woo-woo stuff. This was 2008 at the University of Virginia—a school known more for churning out leaders of the mainstream capitalist economy than eccentric thinkers. Sure, I unfairly judged Permaculture from the get-go, but I already felt like I was pushing the envelope enough just by growing a tomato in my backyard.

Two years later, recently graduated and living in Wyoming, I met another Permaculture zealot named Nick. He was about my age and had picked up the Permaculture bug in his travels. In fact, it was *all he talked about*. We both shared an interest in farming and living close to nature, but I had already prejudged Permaculture, so I wasn't hearing anything he was saying. I just wanted to learn how to grow food and identify plants—I didn't think I needed an entirely new thinking paradigm. After a summer of traveling and camping together out West, we parted ways, Nick heading off to live on a Permaculture farm in Thailand while I hopped on a plane to China to teach English and earn enough money for my next adventure.

Nick and I reconnected a year later to do a month-long trek to Everest base camp in Nepal along with our mutual friend, Barry. We first met up in Bangkok along the infamous Khaosan Road, a bustling commercial street filled with backpackers, cheap hostels, and dive bars. Even among all the travelers from around the world, Nick stood out with his long unkempt hair and beard. While I had spent the last year living in a city of 15 million people, learning Chinese, and losing all the muscle and fitness I had developed in Wyoming, he had spent the year working outside, growing his own food, and learning about subtropical Permaculture. Despite the contrast between us, we quickly fell back into our old camaraderie, the one built on our shared interest in plants and the natural world

We flew from Bangkok to Kathmandu in early October, inadvertently arriving at the beginning of Dashain Festival, one of the biggest Hindu holidays of the year. We wanted to hike the long way to Everest by starting in the foothills instead of taking the common shortcut of flying to the Lukla airport higher up in the mountains. The problem was it was impossible to get a bus ticket because all the locals were heading back to their home villages for the holiday. After hanging around the bus station for several days trying to get a ticket, we finally bribed our way onto a bus, but the catch was that two of us would have to ride on the roof with about 30 other young guys. Being the less brave one in the group and using my "bad back" as an excuse, I claimed the one proper bus seat while Nick and Barry climbed onto the roof. Five bumpy hours later, we got dropped off on a terraced hillside with our gear and were told to "follow that trail" to Everest.

Hiking through the subtropical foothills of northeast Nepal turned out to be one of the most interesting parts of the trek. We walked day after day through rolling hills dotted with small homestead farms and the occasional village. At the end of the third day, we came over a hill to find a small house built on the bench of a slope surrounded by several acres of dense foliage and greenery. It stood out to me because there were more trees planted around it than most of the homesteads. Nick was ahead of me, peering over a stone wall into the yard of the house, waving and yelling, "Check it out!" As I got closer, the details of the garden came into view—lush fruit trees spread out around the house surrounded by small plots of vegetables and shrubs. Gourds and pole beans grew up the trees and hung from the branches while below everything a flock of chickens roamed around picking at fallen morsels of fruit. Nick, wide-eyed and grinning, turned to me and said, "This is what I've been talking about man! This is Permaculture!"

The homestead, it turned out, was also a guesthouse, and we stayed there that night and met the family who tended to the beautiful forest garden. The father harvested a chicken from the yard, and we chatted and drank tea as he effortlessly butchered and broke down the bird for our meal. We dined on chicken, curried potatoes, yellow squash, and lima beans the size of silver dollars, plus heaps of homemade fermented green chile sauce, all from the gardens. Reflecting on our bountiful meal as I hiked the next day, I realized how much the conventional organic farming I knew was missing. Where was the ecological diversity, the integration, and the aesthetic beauty of the forest garden I had seen the night before? "I guess I'll have to look more into this Permaculture thing after all," I begrudgingly admitted to Nick.

Since my Food Forest epiphany in Nepal, I have learned a lot more about the long history of Food Forests, as well as the short history of Permaculture. I learned that variations of forest gardening—or integrated perennial-based horticulture—have been developed many times over by Indigenous people around the world as part of their cultural and ecological systems. I learned about the Hawaiian *ahupua'a* system, the Mesoamerican *milpas*, and the ancient Food Forests of Morocco that inspired Permaculture pioneer Geoff Lawton.[1] When I returned home, I learned that the Indigenous people of Virginia practiced their own form of forest gardening for centuries through intensive management of the woodlands for tree crops and hunting habitat.

Our Approach to Food Forests

The Food Forest ideal was our gateway into our orcharding obsession. Like many others before us, we dreamt of a diverse, low-maintenance forest ecosystem yielding delicious fruit, edible and medicinal plants, and habitat for wildlife and humans alike. For over three collective decades, we have been experimenting—planting trees and shrubs, learning from failures, and replicating successes. In one way, we will never even scratch the surface of what our Indigenous ancestors knew, but we are grateful for the inspiration they have passed down to us. On the other hand, we are lucky to have learned a lot by copying and building on the strategies of recent pioneering practitioners such as Robert Hart, Martin Crawford, Dave Jacke, Stefan Sobkowiak, Michael Phillips, Michael Judd, and Dave O'Neill.

This is *not* a book about Food Forests. This is a book about plants and the relationships we build with them. But plants do not exist alone, they exist as part of a mutually interdependent community (just like us!), and to understand them deeply, we need to engage with that community as a whole. Cultivating a Food Forest, we've discovered, is a wonderful way to grow plants that honors the interconnectedness of nature. It is a way to dance with our plant friends not as solo partners in an open field but as an assemblage of neighbors in a diverse, healthy forest.

Sounds great, right?! Well, it is *and* it turns out it is complex and sometimes very challenging. How to do it successfully depends on a head-spinning amount of variables including your goals, your climate and soil, your budget, and your scale. In this section, we would like to share a basic framework for developing a Food Forest, one that has worked for us and that we think is a good way to get started. It will also serve as a growing pattern that we will reference as we talk about each plant in the chapters to come. Since this is not a book about Food Forests, there are a lot of nuances we won't get to, but we will point you to other resources for more detailed information.

Our goal here is to simplify the Food Forest concept for you, the same way we do in our courses, and to give you the framework and confidence to get started. We also share a list of plants and cultivars that have worked best in our Mid-Atlantic orchards—what we call a "Food Forest Cheat Sheet." One other note: we use the term Food Forest and orchard interchangeably, but just know that when we talk about orcharding, we are talking about diverse, multilayered forest gardening, not modern conventional orcharding that tends to be a monocrop of one tree. For a much deeper dive into designing, planning, and setting up a Food Forest, we recommend checking out *The Food Forest Handbook* by Darrell Frey and Michelle Czolba.

Permaculture Plant Guilds

Guilds are a collection of plants that work together to increase resilience and productivity in a given space. The concept came out of the observation that plants, when left on their own in an ecosystem, tend to collect themselves into mutually beneficial polycultures. A perfect example that we often see in our bioregion is found along many forest edges: oak and hickory trees fill the canopy, small shade-tolerant trees such as serviceberry and redbud grow in the understory, Himalayan blackberry and black raspberry spread out beneath the trees, and wild grape grows up the trees reaching out for the sun. An array of herbaceous perennials thrives under and around the berries, taking advantage of the protection from deer and mowing equipment. Dozens of species flourish together in a small space without any intervention from humans.

The guild concept, like the Food Forest, is an attempt to mimic a successful pattern in nature and simplify it into a useful pattern for growing plants. A lot has been written about guilds since Bill Mollison and David Holmgren introduced the idea, and we are wary of simply recycling other people's ideas. We have even gone back and forth over the years about the actual utility and effectiveness of plant guilds in design. Is it just a shiny idea or does it have any practical value?

But after years testing out guilds on our own farms and engaging in long conversations in our classes, this is where we have landed as a team. Guilds are the perfect design framework to organize a Food Forest around, big or small, as long as we think about them as *useful patterns* and not scientifically proven, set-in-stone plant assemblages. We also advocate a simplified guild framework that can fit almost any orchard size, shape, or layout and that can be scaled for small or large projects.

Our simplified guild pattern contains one anchor plant (usually a fruit tree) in the center of the guild with an array of support plants assembled around it. To maximize the utilization of sunlight in the guild, we stack the plants in three vertical layers: the tree canopy layer, the shrub layer, and the herbaceous layer. Many guild and Food Forest frameworks contain six or more layers, including overstory, understory, vines, roots, fungi, and more. Depending on your goals and scale, we encourage you to play around with these additional layers (and maybe

Food Forests can be designed with multiple layers, including the six shown here. We recommend starting with three layers: tree canopy layer, shrub layer, and herbaceous layer.

even come up with your own!), but we personally find that three is the ideal number to design for a functional orchard that is not overly complex and can be scaled up.

Next, we break the plants in the guild into five different parts based on their function in the guild: anchor plants, nitrogen-fixers, barrier plants, beneficial attractors, and dynamic accumulators. Again, additional categories are used by other Permaculture designers, but these are the five we find most useful at a conceptual level.

The design goal is to have at least one plant from each category in each guild, and to try and select plants that each fulfill more than one role. This design strategy covers the Permaculture principle of "integrate rather than segregate," which our mentor Dave Jacke expanded to mean the following: each element in the design should perform multiple functions (stacking functions), and each function in the design is performed by multiple elements (redundancy).

The five categories, or roles, of a guild are not based on hard science, nor do they need to be for our purposes. They are patterns we utilize to ensure we have diversity in our system—a way to work towards that ideal of functional interconnectedness over time. When we first started designing guilds, we spent hours researching plant functions, doing needs and yields analyses, trying to design the "perfect" guild. At some point, we realized that the whole exercise, and nature itself, was too complex to design, and we just started throwing polycultures of plants together in the orchard to see what worked well.

We realized that there is no such thing as an ideal guild. What is more important is just getting diversity into the system and utilizing light properly to maximize yield. If you are going to be meticulous about something, focus on proper plant spacing and vertical layering, but don't stress too much about guilds. Instead, we encourage you to have fun, experiment, and learn what works on your site as you grow, *and* beware—like just about every permaculture friend we have, both of us have made mistakes. For example, Ryan has 4,000 square feet of mint that he has no idea what to do with. The patch started as one plant divided from the original spearmint plant his grandma used in her sun tea recipe. He's attached but not thrilled about now being the proud owner of what amounts to enough mint to freshen up a sewage plant. Meanwhile his grandma grew it in the windowsill—some patterns don't need to be reinvented.

The plant stories in this book are arranged according to guild function—five of our favorite plants for each of the five guilds: anchor plants, nitrogen-fixers, barrier plants, beneficial attractors, and dynamic accumulators. These five functional categories are somewhat loose and arbitrary, and we encourage you to play around with them and consider creating your own. As you read our stories about plants, friendships, losses, and adventures, we encourage you to extend the guild metaphor to the Human Sector. How do you and your community create a guild? Who plays what role, or function, in your social ecosystem? Like some metaphors, it can seem a little corny, but we think it gets at some larger truths, or at least helps clarify them.

Towards the end of our classes, we ask each design team to get together and reflect on everyone's role in the collaborative design they have created, and then use the guild metaphor to describe each person and their contribution. I remember in my first SPI class as a teaching apprentice, I was described as a mushroom dynamic accumulator. "Trevor operates quietly under the surface," my team said, "collecting and synthesizing information, eventually fruiting out like a mushroom and sharing what he has learned with others." The metaphor helped me think about my strengths and weaknesses in a fun, non-triggering way. I don't know if plants like metaphors or not, but we find them very helpful for understanding our botanical friends, and practical as a tool for moving beyond analysis paralysis.

Anchor Plants

The anchor plant is the central plant that the guild is built around. Usually, it is a tree that you are growing primarily for a specific yield, and sometimes it is considered the most important or valuable plant in a polyculture. The anchor plant doesn't have to be a tree either—you could just as easily design a guild around a shrub or even an herbaceous plant. At Wild Rose Orchard, our anchor plants are fruit trees: apples, pears, persimmons, peaches, plums, and cherries. We also have nut tree guilds in our silvopasture area that include chestnut, black walnut, heartnut, butternut, and hickory. Along the edge of the farm is a timber/firewood guild with black locust as the anchor plant, and in our chicken yard, we have planted forage guilds centered around hazel, Che, staghorn sumac, and cornelian cherry.

Nitrogen-Fixers
Nitrogen-fixers are any plant that can partner with microbes to pull nitrogen out of the atmosphere and add it to the soil. The most common nitrogen-fixers are in the legume family—think beans and peas—but there are many other lesser-known nitrogen-fixers, including many trees and shrubs such as black locust, honey locust, alders, and Elaeagnus species. The idea behind nitrogen-fixers is that they provide natural nitrogen to the rest of the guild, especially when they are cut back, or chopped down, at which point they will slough off roots and release nitrogen into the soil.

Now, depending on your soil and what you are growing, nitrogen might not be a limiting factor in your system, so nitrogen-fixers vary in importance. But our approach to nitrogen fixers—again, not scientific but more experiential—is that they seem to help and certainly can't hurt. One study showed that nitrogen-fixing plants intercropped with wheat had a positive impact on the growth of the wheat.[2] Regardless, they are awesome plants with multiple functions that we want to grow anyway!

Barrier Plants
A barrier plant is any plant—tree, shrub, or herbaceous—that can block out other plants, especially grasses and weeds, from encroaching into an area. The idea is to ring the outer edge of the guild with barrier plants to protect the anchor plant from encroaching weeds that might compete for soil nutrients, water, and light. When trees are getting established, they really do get stunted if grasses grow up around them. To keep a grass-free zone around the tree, we can use mulch (aged wood chip mulch is preferred), but we can also use barrier plants. The best barrier plants grow in tight clumps and spread out vegetatively over the years. Some common barrier plants include daffodils, yarrow, and comfrey.

Beneficial Attractors

Beneficial attractors are plants whose flowers tend to attract many beneficial insects, which are insects that prey on pests in the orchard, thus minimizing the need for pesticides or other pest control methods. The classic example is yarrow—its hundreds of tiny white flowers attract an array of small parasitic wasps that are a natural predator of many common garden pests. Have you ever seen a tomato hornworm sitting dead on a tomato plant with dozens of white eggs protruding from its body? That is the work of a parasitic wasp that has parasitized the hornworm's body as a place to lay its eggs! Beneficial attractors are usually herbaceous perennials and can include many of our favorite ornamental flowers, thus stacking functions for natural pest control *and* beauty.

Dynamic Accumulators

Dynamic accumulators are plants that are thought to go deep into the soil and "mine" nutrients that they then concentrate in their leaves and eventually feed into the topsoil. We can then speed up this process of bringing the nutrients to the topsoil by chopping and dropping the plants, using the leaves to mulch the rest of the guild. Good dynamic accumulators have deep roots, grow quickly, and can be "chopped and dropped" several times throughout the season without harming the plant.

Comfrey is often considered the ideal dynamic accumulator, and it can indeed be chopped and dropped many times in one year. Each time it quickly regrows its leaves. (Side note: most of our favorite Permaculture plants fit into more than one guild category; comfrey is a dynamic accumulator, a barrier plant, and a beneficial attractor.)

We aren't scientists and have not read all the literature on the topic, but it does seem like more research needs to be done on what plants accumulate what nutrients, the mechanisms involved, and how or even if those nutrients become bioavailable to other plants in the guild. One interesting on-farm study in New York State seemed to confirm that many of the plants commonly considered to be dynamic accumulators by Permaculture designers do at least bioaccumulate high amounts of certain nutrients.[3] In particular, this study indicated that the vaunted comfrey does seem to accumulate higher levels of potassium and silicon than other plants. Nonetheless, it would be great to see more research done on this topic.

Trevor's Story of Growing Wild Rose Orchard

Before we move on to the nitty-gritty of how to design and implement a guild-based Food Forest, I want to set the stage with the story of my family's now six-year adventure growing a Permaculture orchard. Through this I hope to show that tending a diverse orchard is, as the saying goes, all about the journey not the destination. The best metaphor I can think of is raising children. No matter how many books you read, you don't really learn until you do it. It's a long process with uncertain and evolving timelines, and there are many joyous peaks along the way, as well as no shortage of challenging valleys.

You can't control it all, but you are an active participant—you can observe, interact, observe, repeat. If you get too caught up in doing it perfectly, or getting to the next checkpoint, you'll miss so much of the magic. Slow down, I often remind myself, be in the moment, enjoy the ever-changing beauty of it all, learn *with* the plants (and the kiddos!) as we all grow together.

How we landed on our farm in Mount Sidney, Virginia, is a long story that involves a mental breakdown, years of not giving up, and some eleventh-hour good fortune. But that is a story for another time. Suffice it to say, in May of 2018, Jenna and I found ourselves broke and pregnant living on 12 acres of beautiful rolling land with a cozy 1,000-square-foot house. It was the most exciting time in our lives.

I was itching to get started throwing plants in the ground, but fortunately I had learned by then to slow down, make a design, and grow gradually. Not only is this a central principle of Permaculture design, which I had been studying for years, but I had also learned by watching other people scale up their farms too fast and burn out. Hell, I had just crashed and burned myself on another project and was still picking up the pieces. It also helped that we had run out of money to spend on the farm.

So, I put aside my dream to start a market farm, and instead Jenna and I created a goals statement and sketched out a loose plan to get there. Our long-term vision, or goals statement, emerged: "Wild Rose Orchard is a perennial fruit farm and nursery that unites the wild and cultivated, connects people to the wonder of nature, and nourishes our community with an abundance of healthy food."

Our plan would be to start with just a half-acre diversified orchard, or Food Forest, right outside our house. In this orchard, we would plant as many promising cultivars of perennial fruiting plants as we could afford to buy, test out what grows well on our land, and discover what our family enjoys eating. This "experimental orchard," we envisioned, would eventually serve as the genetic stock for a small nursery business, as well as larger-scale plantings on the rest of the farm.

Every spring and fall since our first son was born, we put in a round of plants. Every summer, my rock star father-in-law, Joel, weeded the orchard while Jenna and I worked off-farm jobs and tended to the kids. (If I have any secret weapon to speak of, it's Joel!) It hasn't always been easy. Our lower field flooded. The chimney blew off our roof. The deer got through our fence and ate our blueberries. We wasted a whole season growing CBD hemp. And we've had more than a normal amount of illness and loss in the family over a short period of time, which has made the farm seem less of a priority.

But the great thing about perennials is that when the humans step away for a bit, they pretty much just keep doing their thing. Six years later, our half-acre Food Forest is finally planted out and contains 15 species and 80 cultivars of fruit trees, dozens of fruiting shrubs, and hundreds of herbaceous perennials. Jenna and I still work full-time jobs off the farm, and we now have two children, but Wild Rose Orchard continues to slowly evolve towards our original vision.

Throughout all of this, our orchard and berry patches have been a central part of our family's daily and seasonal rhythm. Every night after dinner, we walk the farm together. During the summer, this walk turns into a berry-picking marathon, with the challenge always being to pick more than we can eat so we will have enough left to freeze for the winter.

In year six, the orchard is in what I call the "adolescent" stage. It's rapidly changing and volatile. The greater orchard ecosystem is starting to form and synergize, but it's not quite taken off yet. New types and varieties of fruit are coming online every year, which is exciting. Last year all of the tree fruits got wiped out by a late frost, but the berries rocked out. This season has been a banner year for tree fruit—apples, peaches, plums, cherries, pears—and we are getting to try over a dozen varieties that we've never tasted before.

Last night was one of those peak farm nights. Before dinner was even over, my kids were begging to go out into the orchard and pick goumi berries. When we got out there, the late afternoon summer sun was still roasting hot, but luckily the goumis were sitting in the shade of a five-year-old peach tree, now 15 feet tall and loaded with fruit. The kids raced to the goumis, and instead of moving around the orchard checking on all of the plants like I usually do, I remembered the Permaculture principle, "obtain a yield."

So, I sauntered up to the kids and sat cross-legged in the shade right in front of my favorite gooseberry cultivar, "Jeanine." Slowly and deliberately, I started snacking on the delicious berries. I savored each one like the work of art it is, riding the gastronomic wave of the underrated gooseberry. First the crunch and the burst of tart juice, then sweet velvet flesh, then delightful chewy skin to round it off. The perfect dessert.

I took a deep breath, straightened my back, and took in the miraculous scene in front of me as I slowly exhaled. Noemi, my three-year-old daughter, was in the zone, silently picking and eating goumis from the lower branches of the tall shrub. My six-year-old son, Afton, had already moved on to other things, roving around the orchard checking on the other fruits, all the while reporting back excitedly.

"These peaches are almost ripe Dad!"

"There's a spider on the elderberries!"

"These black currants are hitting hard!"

Jenna meanwhile sat in peace on the front porch reading, happy to have a few minutes of solo time in the shade, occasionally glancing up and smiling at her wild little family.

These are the perfect moments that emerge out of the complexity of the living farm system and make all of the hard work and sacrifice worth it.

Okay, back to the task at hand, now where were we? Ahhh yes, Food Forests, site selection and layout …

Site Selection and Layout

Now that we have laid out the framework for what a guild is, let's talk about how to use the guild pattern to design and implement a Food Forest or orchard. The basic idea, which we flesh out below, is to put multiple guilds together in a repeating pattern to create an orchard. Before we dive into that, let's talk about selecting the size and location of the orchard.

The size of your orchard comes down to your individual goals and the time and resources available to you. As with everything in Permaculture, we always recommend starting small with a pilot project. Try planting two or three guilds at first and spend a few seasons working with these pilot guilds to get a feel for the time and labor required. Then, create a well-thought-out design for your larger Food Forest based on your goals, research, and this initial personal experience.

The size of the orchard you create will affect a lot of your design decisions. In general, the larger your orchard is the more you will want to simplify your design to keep access, harvest, and maintenance manageable. For example, a small orchard with three fruit tree guilds is very easy to manage in a hands-on way. If certain plants start to get out of control or take over, you can go in yourself and cut things back by hand in no time. Similarly, any chop and dropping, mulching, or holistic spraying you need to do is a small task. On the other hand, a five-acre orchard with a thousand fruit tree guilds is too large for one person to manage by hand, so you will want to simplify to make maintenance as easy as possible, and you will want to design in access for machinery and/or livestock.

For a homestead-scale orchard that is mainly for personal use, we recommend anywhere from a small plot of three tree guilds to one acre of tree guilds. For reference, Wild Rose Orchard is half an acre and contains 80 fruit tree guilds. It is just big enough for our family to have plenty of fruit throughout the summer with a little bit of surplus to share with friends and neighbors. It is also small enough for two people to manage mostly by themselves on a very part-time basis with just a riding mower, tow-behind dump cart, and a battery-powered backpack sprayer. If it was much larger, it would become unwieldy without larger mowing, mulching, and spraying equipment and/or additional labor.

Selecting the exact location and layout of your orchard will be influenced again by the size of the orchard as well as the spaces you have available to you. In general, fruit trees like to grow on gently sloped ground or a hillside that allows for excess water and cold air to drain away. But don't be deterred from growing on flat or low-lying ground, especially if you are starting small and expanding over time to see what works.

There are two types of Food Forest layouts that we recommend, each of which we will delve into below: the linear guild Food Forest and the expanding island Food Forest. In general, we find the linear guild layout appropriate for orchards on hillsides and larger orchards (say, over a quarter-acre). We like to use expanding island guilds for small orchards (less than a quarter-acre) on flat ground.

Linear Guilds

The linear guild Food Forest is made up of tree guild rows that run parallel to each other and are spaced out evenly across the landscape. For example, Wild Rose Orchard is made up of parallel linear rows spaced

The half-acre experimental Food Forest at Wild Rose Orchard in Mount Sidney, Virginia
CREDIT: SCOTT TURNER

30 feet apart. The space between the rows, or the alleys, can be kept open as grass, grazed by animals, or used for planting low-growing annual vegetables or perennials like asparagus.

Within the row, or linear guild, we place our guild plants in a repeatable pattern to get the type of vertical layering we want. In our orchard, we use a pattern of fruit trees with two shrubs in between each tree and an herbaceous perennial plant in between each shrub/tree. The pattern you choose will depend on your spacing, the size of the plants you are using, and how complex you want to make things, but we recommend keeping the pattern simple and repeatable.

The linear guild Food Forest was first brought to our attention by our friends at Radical Roots Farm in Keezletown, Virginia. Dave and Lee O'Neill of Radical Roots have laid out their entire five-acre vegetable farm in a linear guild pattern. In their case, they spaced their guild rows very far from each other so that they could keep the alleys in full sun for their vegetable production, which is their main cash crop.

Dave and Lee's farm was the first production system we saw laid out in a linear guild framework, but we have since learned from several

In this linear guild layout, two shrubs and three herbaceous plants are planted in between each tree.

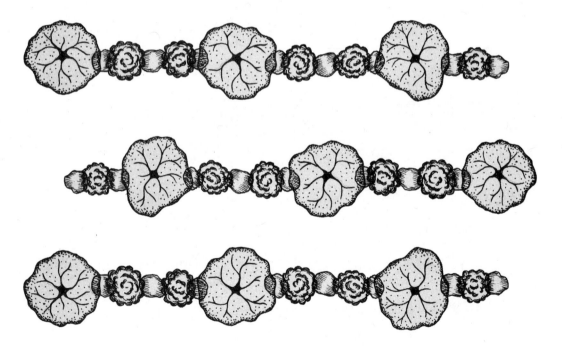

other pioneering farmers. Stefan Sobkowiak's 5-acre Miracle Farm in Quebec is one of the best examples we have seen of a commercial U-Pick orchard laid out in linear guilds. A little larger, and on the other side of the Atlantic, Richard Perkins' 25-acre Ridgedale Permaculture farm in Sweden utilizes the linear guild idea as the organizing pattern for its many orchards and silvopasture blocks. Both of these are great resources if you are interested in setting up a linear guild orchard.

One potentially tricky thing about laying out a linear guild on a slope is that it is ideal to keep the alleys between rows a consistent width—30 feet, for example—to make mowing or grazing in the alleys efficient and uniform. The easiest way to keep the alley widths uniform is to lay the orchard out in a "keyline" pattern, a strategy developed by P.A. Yeomans in Australia. Keyline design is an entire design strategy in itself, but what we want to glean from it for our orchards is the specific technique of using a keyline pattern to lay out our linear guild rows.

Be warned, keyline patterning can seem complicated and confusing at first. It is also notoriously difficult to teach. You kind of just have to try it out on a small scale until it clicks. There are also many different ways to arrive at keyline patterning, and a lot of debate about which is the correct way. True keyline patterning will actually move water passively from the valleys to the ridges. However, for our purposes, we are just trying to lay out rows that are uniformly parallel to each other. If you are digitally inclined and good with contour maps, you can lay these lines out on a digital map and then use a GPS receiver to find and mark the points on the ground. The agroforestry software Overyield has a built-in keyline tool that also works well for creating parallel planting rows. If you are working on a small scale, you can simply find one contour line in the middle of your Food Forest and mark parallel rows uphill and downhill from that point.

Linear Guild Spacing

So, what is the best spacing for plants in a linear guild? This of course depends on what you are growing. But in general, we are going for wide spacing to make sure sunlight continues to penetrate the understory of the guild even when the trees are fully grown. In this sense, we are not actually trying to replicate a dense, shaded forest, but rather an early- to mid-succession forest or savannah. This type of open, semi-shaded system is best at partitioning sunlight to all the plants and also allows for airflow, which is particularly important in humid climates like the Mid-Atlantic.

A good rule of thumb for spacing within the row was developed by Martin Crawford in his excellent book, *Creating a Forest Garden*. He recommends a minimum spacing between trees of 50% of the mature diameter of the tree in between each tree when fully grown. This can most easily be explained in the diagram below. For example, a row of

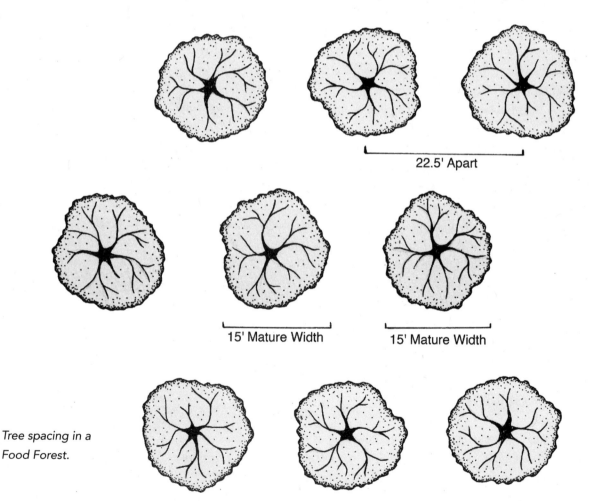

Tree spacing in a Food Forest.

trees that grow to 15 feet wide at maturity should be planted at least 22.5 feet apart within the row.

Spacing between each row, or linear guild, should be at least as wide as Martin Crawford's rule of thumb, and often even wider depending on what you plan to use the alleys for. The alleys at Wild Rose Orchard are 30 feet wide, which leaves a nice open area for mowing and walking. If you plan to graze in between the alleys, you will probably want to go wider—at least 40 feet for sheep and 60 feet for cows.

For the shrubs and herbaceous plants in a linear guild, how much you stack in between each tree depends on how wide you space the trees. In Wild Rose Orchard, the trees are planted 22 feet within the row, and we have found that there is enough space between each tree for two small shrubs and three herbaceous perennials or one large shrub and four herbaceous perennials. This pattern of Tree - Herb - Shrub - Herb - Shrub - Herb - Tree then becomes the guild itself. To complete the design, string together as many guilds as needed to fill out each row. Below, we will share tips on how to simplify this repetition even further based on your scale and goals.

Expanding Island Guilds

The expanding island guild, first demonstrated to us by Michael Judd, is most suitable for small flat spaces. It involves creating circular guilds spread out evenly from each other like islands in a sea of grass. Support plants are assembled around the anchor tree and can expand over time as you plant and propagate more.

Spacing between trees is straightforward—simply follow Martin Crawford's rule of thumb of a minimum spacing of mature tree width plus 50%. The space in between island guilds can remain as grass and be mowed as an access path, or it can be expanded over time as you grow your guilds.

Spacing and arrangement within the island guild is a little trickier than a linear guild and depends very much on the size and nature of the plants themselves. In general, we recommend experimenting with different arrangements in your own pilot plot, but there are a few guidelines

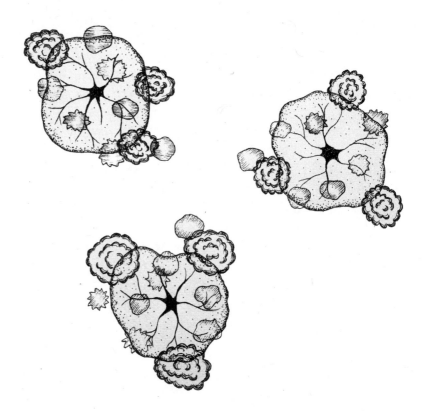

Expanding island guild layout.

we have found helpful over the years. For one, try and put shrubs on the outer edge of the island to make them easy to access from the grass paths and to prevent them from crowding the base and lower branches of the tree. Also, place sun-loving plants like goumi and elderberry on the south half of the island and true shade-tolerant plants like currants and gooseberries on the north half. Quick-spreading plants like comfrey and mint can be planted on the northern shady side of the trees to keep them in check. (These north-south recommendations are for the temperate latitudes in the Northern Hemisphere and would be the opposite in the Southern Hemisphere).

What Is a Permaculture Plant?

Before we jump into the fun stuff—selecting the best plants and cultivars for our guilds—we'd like to share our definition of a Permaculture plant. To us, a good Permaculture plant, meaning a plant we would want to consider putting in our systems, must meet at least one and ideally all three of the following criteria.

1. Multifunctional: Does it perform more than one function in our system other than being edible or medicinal? Can it also be used for livestock, or does it perhaps fit into one or more guild roles? A perfect example is black walnut, which provides an edible and medicinal yield, valuable hardwood timber, and habitat for wildlife.
2. Low maintenance: Is it easy to care for with minimal management and low to no sprays? A good example in our climate would be the resilient Asian pear.
3. Native when possible: Although we are not dogmatic about it, we strive to include as many plants native to our bioregion as possible. These plants tend to be resilient to macro- and microclimatic changes. Native trees like pawpaw and persimmon are adapted to our erratic spring temperatures and late spring frosts and do not lose their fruit to frost kill as often as non-native fruits. Native plants in general provide habitat and a food source to abundant native wildlife that we rely on to keep our cultivated Food Forest ecosystems balanced.

Plant Selection

Now that we have some basic criteria for choosing the Permaculture plants in our guilds, it's time to start selecting a layout of species and cultivars. Let's say you are designing a half-acre linear guild Food Forest; how do you decide the planting and guild sequencing? On a small scale, you could throw in a high degree of variability and observe what works well, but if you're planting half an acre you might want a general pattern to repeat.

One approach developed by Stefan Sobkowiak, called the "Grocery Aisle" concept, is to group plants that generally ripen around the same time together in each row. For example, Row 1 at the top of the hill might contain the earliest blooming and ripening fruits, which helps them avoid late frosts that will sink to the bottom of the hill. As you move down the hillside, each row will contain a grouping of plants that bloom a little bit later. The Cheat Sheet below has been arranged, very approximately, to group tree species and cultivars together based on Stefan's Grocery Aisle concept.

Another approach is to pick one alternating guild pattern per row. For example, Row 1 might contain all Asian pears for the tree layer and all black currants for the shrub layer. The advantage of this is that it makes management and harvest easier because you can focus your efforts in each row on the needs and timing of a single species. You still get diversity *within* the orchard because the next row down will contain a different grouping of plants, for example, persimmons and elderberries. Wild Rose Orchard is laid out using the Grocery Aisle approach, but if we were to design a commercial orchard, we would probably use this more uniform approach. Also, if you do choose this approach, we would still recommend maintaining a lot of diversity in the herbaceous layer.

The species and cultivars listed below have a high potential for pest and disease resistance in the Mid-Atlantic. This is not an exhaustive list, but it is a starting point for some of the most promising cultivars available. Please note that there are many other wonderful perennial plants that are not included in this list because they have very specific site or management requirements that make them more suitable to being planted in blocks by themselves. For example, we recommend planting pawpaws together in a large patch because they struggle with pollination and like partial shade during establishment. They can certainly be incorporated into a Food Forest if you'd like, but depending on your context, that might not be the best location for them in your system. Similarly, cane fruits (raspberries, blackberries, etc.) are best grown in linear rows where trellising and harvest can be efficiently carried out. Other trees and berries that we recommend planting in standalone blocks or rows include blueberry, seaberry, hardy kiwi, grapes, and figs.

The Food Forest Cheat Sheet
The Tree Layer
Early/Mid-Summer Harvest
- Peaches: Burbank July Elberta, Early White Giant, Blushing Star, Intrepid, Contender, Coral Star, Indian Free, Buenos II, PF 13 Lucky
- Sour Cherries: North Star, Danube, Surefire
- Sweet Cherries: White Gold, Black Gold, Stella
- Plums: AU Roadside, Shiro, AU Rosa
- Apples: William's Pride, Pristine, Dayton
- European Pear: Ayer's, Moonglow, Gem, Warren
- Mulberry: Trader, Oscar, Miss Kim, Illinois Everbearing, Silk Hope, Gerardi (dwarf), Beautiful Day, Varaha, Taylor #1, Madhava

Late Summer Harvest
- Apples: Empire, Priscilla, Pixie Crunch
- European Pears: Blake's Pride, Harrow's Delight, Maxine, Spalding
- Asian Pears: Shinseki, Hosui, Kosui
- Beach Plum: Premier, Jersey

Fall Harvest
- Apples: Liberty, Freedom, Enterprise, Goldrush, Arkansas Black, Sundance
- European Pears: Shenandoah, Potomac, Magness, Kieffer
- Asian Pears: Shinko, Chojuro, Korean Giant, Shin Li, Daisui Li, Kikusui, Yoinashi
- Asian Persimmons: Nikita's Gift, Ichi Ki Kei Jiro, Hana Fuyu, Jiro, Matsumoto, Saijo, Olympic
- American Persimmon (90-chromosome type): Prok, Early Golden, Geneva Red, King Crimson, Mackenzie Corner, H63A, 100-46
- Che (seedless): California Dreaming, Norris, Darrow, Adams
- Jujube: Li, Honeyjar, Sugar Cane, Shan Xi Li, Coco, Black Sea

Tree Rootstock Recommendations

In general, we recommend semi-dwarf rootstock. Dwarf rootstocks will bear fruit sooner but are more fragile and shorter-lived. Standard rootstocks are strong and long-lived but take many years to bear fruit. We have found that a semi-dwarf rootstock such as the M7 for apples is a nice happy medium. M111 rootstock is also a very popular rootstock and produces a fairly large tree about ⅔ standard.

The Fruiting Shrub Layer
- Small Shrubs
  - Honeyberry: Aurora, Boreal Beauty, Boreal Beast, Boreal Blizzard, Indigo Gem
  - Gooseberry: Jeanine, Invicta, Captivator
  - Jostaberry
  - Black Currant: Titania, Otelo, Risager, Whistler, Blackcomb, Black September
- Large Shrubs
  - Goumi: Sweet Scarlett, Red Gem
  - Elderberry: Ranch, Marge, Bob Gordon, Wyldewood, Pocahontas
  - Nanking Cherry
  - Aronia, aka Chokeberry: Viking, Nero, McKenzie
  - Seaberry, aka Sea Buckthorn

Nitrogen-Fixers
- False Indigo Bush, aka Indigo Bush (*Amorpha fruticosa*)
- Smooth Alder (*Alnus serrulata*)
- Siberian Pea Shrub (*Caragana arborescens*)
- Goumi
- Seaberry, aka Sea Buckthorn
- Black Locust (*Robinia pseudoacacia*)
- Redbud (*Cercis canadensis*)
- Blue False Indigo (*Baptisia australis*)
- Clovers

The Herbaceous Layer
- Yarrow (*Achillea millefolium*)
- Bee Balm (*Monarda* spp.)
- Mountain Mint (*Pycnanthemum* spp.)
- New England Aster (*Symphyotrichum novae-angliae*)
- Echinacea (*Echinacea purpurea*)
- Blue Vervain
- Tulsi, aka Holy Basil
- Angelica (*Angelica archangelica*)
- Sweet Cicely
- Comfrey
- Rhubarb
- Horseradish
- Garlic
- Daylilies, Daffodils, and other spring bulbs
- Sunchoke, aka Jerusalem Artichoke
- Zinnia
- Cannabis
- Strawberry
- Stinging Nettles (*Urtica dioica*)
- Beets
- Burdock (*Arctium lappa*)
- Peppermint, Spearmint, etc.
- Egyptian Walking Onion
- Sea Kale
- Perennial Sorrel
- Creeping Thyme
- Roman Chamomile
- Sorrel
- Asparagus
- Good King Henry
- Turkish Rocket
- Bronze Fennel

Preparation and Planting

When you are ready to plant a tree, make sure you follow proper planting protocols: digging a large planting hole, pruning damaged roots and spreading them out evenly, and not planting too shallow or too deep. Bare-root planting in fall or spring is much more successful than summer planting. I like to dunk the plant roots in a mycorrhizal inoculant slurry before planting. I also amend each tree planting hole with approximately two pounds of azomite (for micronutrients) and two pounds of rock phosphate, mixed evenly into the soil.

Once trees are planted, they will need an average of one inch of rain per week to get established. They also need a weed-free zone maintained around the drip line at all times. Dwarf trees will need to be tied to a stake. The cheapest option I've found is a three-quarter-inch electrical conduit. Sink the stake two inches from the trunk of the tree on the upwind side and then fasten the tree to the stake using a flexible tree lock.

A tree guard placed around the trunk of the tree at the base is essential for preventing rabbits and voles from girdling the tree, and branch tips must be protected from deer browse. Proper training from the outset is important to minimize corrective pruning later on. Seasonal pruning is recommended to encourage proper airflow and branch distribution and maximize fruiting.

Pest and Disease Management

There's no doubt that growing fruit trees organically is a challenge, especially in humid temperate climates. We've already gotten part of the way there by choosing disease-resistant cultivars and creating diversity in the Food Forest. From there you can choose to target problems with certified organic sprays or experiment with probiotic and holistic sprays to boost the plant's ability to fight off pests and diseases. To learn more about holistic sprays, we highly recommend Michael Phillips' book, *The Holistic Orchard*. Michael developed a set of holistic sprays that worked well on his orchard in New Hampshire and that he was still working on improving when he sadly passed away in 2022. Michael's work continues to be a major inspiration for us as we learn more every season about how to work with our beloved perennial plants.

If you are interested in experimenting with low-cost DIY holistic sprays, you can also check out Nigel Palmer's book, *The Regenerative Grower's Guide to Garden Amendments*. We also recommend following the work of John Kempf and his company, Advancing Eco-Agriculture, as they are continuously developing better and more efficient methods of boosting natural plant health and defenses using biological inputs and sprays. Other promising approaches to holistic plant health management include JADAM, Korean Natural Farming, and Biodynamics.

Michael Phillips' Holistic Spray[4]

Makes a 4-gallon ready-to-spray batch. Mix in a 5-gallon bucket in this order:

- ⅓ cup neem Oil—warm up to liquefy. Omit if spraying pears.
- 2 tablespoons karanja oil (or ⅜ cup karanja if you're spraying pears and thus omitting neem).
- 1 teaspoon biodegradable liquid dish soap—stir to emulsify the oils.
- 1 quart warm water (around body temperature)—stir again.
- Optional: ¼ cup blackstrap molasses to stimulate beneficial microbes and increase Brix in fruit.
- 1¼ cups fish hydrolysate—stir again.
- ¼ cup Kelp-It Concentrated Liquid Seaweed Extract.
- 1¼ cups activated EM-1 (see recipe above).
- Add water (lukewarm is best) until you reach the 4-gallon mark, then stir one last time.

This is a basic holistic spray recipe that can be used for plant health throughout the season. It can also serve as a base recipe for adding other botanical inputs such as nettles, comfrey, horsetail, etc.

Sources for Plant Materials

Below is a list of some of our favorite plant nurseries and seed companies, with a note about which plants we think they have a particularly good selection of. The specialty/unique offerings listed are subject to change and are just our opinion based on our experience ordering plants, so take it for what it's worth!

Nursery/Seed Company	Website	Specialty/Unique Offerings
Edible Landscaping	ediblelandscaping.com	Seedless Che
Peaceful Heritage	peacefulheritage.com	Pawpaws, Figs, Mulberries, Berries
Hundred Fruit Farm	hundredfruitfarm.com	Pawpaws, Ribes
Perfect Circle	perfectcircle.farm	American Persimmons, Mulberries, Chestnuts
One Green World	onegreenworld.com	Figs
Raintree Nursery	raintreenursery.com	Currants and Gooseberries
Hartmanns	hartmannsplantcompany.com	Hardy Kiwi
Honeyberry USA	honeyberryusa.com	Honeyberry
Nourse Farms	noursefarms.com	Raspberries, Blackberries, and Blueberries
Silver Run	silverrunforestfarm.org	Native and Multifunctional Trees Adapted to the Mid-Atlantic
Twisted Tree Farm	twisted-tree.net	Native and Multifunctional Trees Adapted to the Northeast
Seed and Soil	seedandsoilmaine.com	Cannabis Seeds
Strictly Medicinals	strictlymedicinalseeds.com	Medicinal Plant Seeds
Southern Exposure	southernexposure.com	Vegetable Seeds Adapted to the Mid-Atlantic Region
Richters	richters.com	Herbaceous Perennials

CHAPTER 3

Anchor Plants

1. MULBERRY

Trevor's Story

One of the best things to come out of the Internet in the early aughts was a free website called Couchsurfing.com. I think it still exists, but I've either become too old and jaded to use it anymore or it's been ruined by scammers, or both. The idea is an experiment in social trust and goodwill—you build a profile which allows you to find people around the world who are offering free lodging or a couch to crash on. In return, you are expected to be a host at your place. As you couch surf around with other people and also host guests, you get reviews and build a profile of ratings. The idea is that anyone who is sketchy or dishonest will be weeded out with bad reviews.

For a college kid on a budget with a love for traveling, couch surfing was a game changer, and a great place to meet new people. When I biked across the U.S. with some friends after my sophomore year in college, we couch surfed in every major town we went through. In Boulder, Colorado, halfway through our trip, we had so much fun crashing with some CU Boulder students that we ended up staying for two weeks, exploring the canyons outside of town, relaxing in the park and slacklining all day, and going to jam sessions in the evening. In Portland, Oregon, our host wasn't even there when we arrived, but she left a key under the doormat and a note on the table that said, "Here are the keys to my car, it's parked outside, feel free to use it to explore the city." The level of generosity and blind trust that I encountered in my early couch surfing days blew my mind and buoyed my faith in humanity. Maybe all that shit in the media is wrong, maybe most people are good honest humans after all.

When I finally got a decent apartment for my junior year, I convinced my skeptical roommates to let me list our couch on the site and start hosting. For the most part, it went great, although a few of the encounters did validate my roommates' wariness. The very first guy we hosted lived in his van and quickly made himself at home, taking 45-minute showers twice a day and using our printer to create flyers that he put up around town asking people to send him money in return for good karma. When he wouldn't leave after a week, me and my roommates had to band together and force the issue with some stern words. Then there was the wild group of teenage musicians from Philadelphia who threw an epic concert in the living room but took off the next day with a couple hundred dollars of coins that had been sitting on my roommates' dresser.

Nonetheless, after writing them all negative reviews, we pressed on, meeting all kinds of interesting people along the way. Some of them became friends who I would end up staying with when I traveled in later years. One guy named Neric, a Buddhist mathematician from LA, was the first person to teach me how to do yoga properly after he noticed I always had a sore back. I still do Neric's version of warrior pose regularly.

Then there was the Turkish photographer named Kemal who first introduced me to mulberries and inspired my lifelong obsession with wild foraging. Kemal was a middle-aged guy whose English was not very good, and so he communicated with jubilant facial expressions and gesticulations. He was an accomplished nature photographer who traveled all over the world, and since we couldn't communicate very well, we spent time looking at photography on his computer and going on walks around the neighborhood.

One balmy summer evening, we walked downtown to look at a photography exhibit where his work was being featured. As we approached a busy intersection, Kemal darted across the street into the median strip and started shouting with glee and waving me over as he reached into the branches of a small tree. He was stuffing purple berries into his mouth when I arrived, and he thrust some into my hands yelling the Turkish word for mulberry followed by "EAT!" The mulberries were sweet and delicious, and we went from tree to tree gorging ourselves like ravenous but happy children. After a few minutes, the sky opened up and started drenching us in summer rain, but we just kept on foraging, laughing with glee as the mulberry juice on our faces mixed with the rain and became a purple river running down our bodies.

How to Grow

Mulberries are a diverse genus of trees that include three main species: *Morus rubra*, native to North America; *Morus alba*, native to East Asia; and *Morus nigra*, native to Central Asia. Fortunately for us, all three species hybridize easily, which has led to a wide array of cultivars that can be grown from chilly Zone 4 to subtropical Zone 9. Here in the Mid-Atlantic, many of our cultivars are hybrids of the *Morus rubra x alba* lineage.

Mulberries are sun-loving, fast-growing trees that often show up in abandoned fields and urban lots and are considered by the uninitiated to be junk trees. In the wild, the flavors range from bland to delicious, but the cultivars have been bred to have large sweet fruits. Full-sized trees can eventually be 40 feet tall, and there are also excellent dwarf varieties available that grow to only 10 feet.

Integrating into Your Plant Guilds

Mulberries are a great anchor tree for a Food Forest guild and also have tremendous value as an agroforestry tree, specifically as fodder for livestock. The everbearing varieties that were popular in the American Southeast before industrialization were bred to drop fruit for many weeks and were used as feed for hogs.[1] Mulberry leaves are high in protein and can be used as direct browse fodder for sheep and goats—even cattle will derive nutrition from the fallen leaves. On our farm, we grow select cultivars in the orchard for fresh eating and also plant out seedlings in our agroforestry areas to fill out the canopy quickly with vegetation.

Cultivar Recommendations

Cultivar selection mainly depends on your hardiness zone. Here in Zone 6, we are forced to plant some of the more hardy varieties from North America and East Asia but cannot grow some of the larger fruited varieties from the Indian subcontinent. Until recently, we didn't have to worry much about pests and diseases; however, we are now getting a fungal pathogen called popcorn disease that can destroy the fruits as they form. This disease seems to be moving north into the Mid-Atlantic, although we hear it has not yet made it to the Northeast. Alas, our favorite variety, Illinois Everbearing, is reportedly susceptible. We are now growing varieties that are supposedly immune to popcorn disease such as Trader, Miss Kim, Silk Hope, and Gerardi (dwarf). For warmer climates of Zone 8 and up, Chiang Mai #60 is a very productive cultivar.

Propagation, Harvest, and Uses

Mulberries are easy to grow from seed and can be grafted to maintain or propagate a specific cultivar. They also can be rooted easily from cuttings, unlike a lot of other fruit trees.

In our experience, mulberries are a hit with children and so would make a great addition to a U-Pick operation. The fruit is delicious out of hand as well as in jams, jellies, and sauces, and it freezes well. We like to add mulberries to our cider ferments along with other blue/purple fruits like aronia and honeyberry. In addition, the leaves are used medicinally in teas and the stumps, and branches provide a great substrate for oyster mushroom cultivation.

2. WILLOW

Ryan's Story

Ever received a book only to put it up on the shelf for years, forget about it, and then pull it out bored one day only to discover that it's one of the best books in your library? As a book lover, I can't tell you how many times I've had this experience. It's wonderful really, when the boring becomes exciting, the mundane novel. This is exactly how I feel about willow.

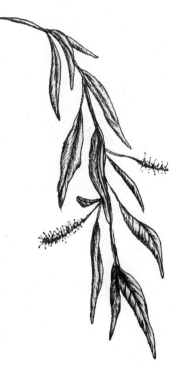

My love of the outdoors may have been seeded in my childhood, even though much of my adolescence was spent in a gym and the last place I wanted to be was outside. My connection to nature was corrected years later in Hawaii where I finally discovered a love of the outside and Permaculture. Surfing every day, hanging out with hippies, and studying poetry will shift your consciousness rather quickly. That time of my life was a time of healing and discovery of all things new and novel for me. It was also a time when I wanted to distance myself as much as possible from what I remembered as the status quo, normal, suburban lifestyle. I rejected everything from my adolescence. The fast food, shitty restaurants and pizza places, the gym, the sneaker culture, video games, air conditioning, showers. Almost everything that reminded me of the convenience and privilege and disconnect from nature of my lifestyle in Virginia, I wanted to throw away. Being a broke twenty-something in a new-to-me culture made this easy.

I found myself engaging plants that were often exotic to me, but native to Hawaii, discovering monkeypod trees, strawberry guava, citrus. As I reflected on my upbringing, all of the plants from my childhood felt so stiff, so tamed. Boxwoods, daylilies, yews, iris. You know the ones I'm talking about. I chalked the "boring" plants up to a cookie-cutter existence, reserved for the well-heeled with a need for symmetrical tidiness and no desire to connect to anything without air conditioning, cable TV, and tchotchkes in the window.

There was, however, one tree from my childhood that I could never shake. This tree, the willow tree, was the wild card. In a landscape full of standstill ho-hum, the willow shimmied.

We moved from a grainy woodsmoke existence just outside of Charlottesville, an area called Greene County, to Waynesboro when I was eight years old. Waynesboro felt like a big city.

My first time exploring the new house and tiny yard, I discovered a giant tree with limbs hanging in a manner creating a dramatic shelter underneath. Immediately, my sister and I gravitated towards the tree to play. I was fond of climbing the tree and swinging from its branches. For the transition period of moving from Greene County to Waynesboro, before we met friends and our social lives blossomed into small-town camaraderie, the tree was our escape.

One day that spring, I got off the bus from school and ran into the house to throw my backpack down and spring back outside to start swinging from the branches of the willow tree. Mid-sprint, I stopped on a dime when I saw only a pile of sawdust where the tree used to stand. The tree was gone. Something about sewer lines and falling branches. My parents dismissed our grieving over dinner that night by sharing the practical problems the willow presented to our sleepy neighborhood. It was replaced by a Leyland cypress. Writing about it now, I can't remember if I bought their explanation or not. The convenience and ease of suburban life beckoned, and my young self easily slipped into the comfortable embrace of video games and Domino's pizza.

But the shimmying unkempt wildness of the willow tree never left my consciousness. Striking a romantic, slightly dangerous pose in my mind, this tree was the seed lying dormant in my psyche waiting for me to discover the plant world years later.

How to Grow

Willow (*Salix* spp.) is so easy to grow that we often recommend it to people as a great starter plant to develop a relationship with. In particular, willow roots easily from cuttings, almost as if it is asking to be propagated far and wide.

That being said, the *Salix* genus is huge and varied, encompassing dozens of varieties from all over the world, some with slightly different growing habits. In general, most willows do well in wet soil, although some prefer more well-drained loamy soil. Because of its tolerance for wet feet, we recommend utilizing willow in areas that are periodically inundated with water—think creek banks, riparian areas, or drainages that flood when it rains.

Integrating into Your Plant Guilds

One useful willow pattern is to incorporate it into a "multi-kingdom remediation" planting anywhere on your property where water flows that you would like to clean before releasing it back into the system. This method was introduced to us by our friend and collaborator Mark Jones of Sharondale Farm.

At Ryan's Dancing Star Farm, a drainage brings water from his neighbors' corn and soy fields onto his property. Ryan was concerned about herbicide runoff from his neighbors' property, so we installed a system of willows and fungi to clean or remediate the water before it flowed onto the rest of his farm. First we took a 20-foot section of the ditch that was carrying the runoff and filled it with burlap sacks stuffed with the sawdust spawn of several saprophytic fungi, including Oysters (*Pleurotis* spp.) and Garden Giants (*Stropharia rugoso-annulata*).

Next, we planted live willow stakes directly into the bags and down into the soil, thus pinning them down. Once the willow rooted and began to grow in the spring, the system evolved into a multi-kingdom living filter—as the water passed through when it rained, it was filtered by the complex microbiology of the system, all the while keeping the plants and fungi hydrated and thriving. This system can be adapted and expanded upon anywhere on your property that receives periodic runoff, and other water-loving plants and fungi can be added in to create even more diversity.

Willow is also very versatile in any kind of Food Forest or agroforestry system because of how easy it is to get established. Especially in large-scale agroforestry systems, we like to include black willow (*Salix nigra*) or hybrid willow (*Salix x matsudana x alba*) because they grow fast and develop into trees well ahead of the slower-growing hardwoods. For example, a healthy 5-foot live willow stake can grow into a 15-foot tall tree within one growing season! Willow is also a popular species in silvopasture systems (integrating livestock like cattle into a tree orchard) because it provides medicinal benefits to the livestock when grazed. Ecologically, willow is one of the few plants that forms symbiotic relationships with ectomycorrhizal fungi, so there is reason to believe that integrating willow into a Food Forest helps increase the mycological diversity in the soil, which is almost always a good thing when you are trying to garden organically.

Harvest, Uses, and Variety Recommendations

The human uses of willow are multifaceted and too numerous to list. Perhaps most famously, willow contains salicylic acid, the active ingredient in aspirin, and therefore has a long history of medicinal use, particularly the white willow (*Salix alba*) in Celtic and Druid traditions. Willow has also been used for millennia to weave baskets and in other building and craft applications—varieties have been bred specifically for these uses. Our favorite varieties for arts and crafts, as well as for adding whimsical beauty to the landscape, include flame willow and corkscrew willow. Due to its powerful growth hormone, we have used it successfully as a rooting agent. Simply stick a small piece of pruned new growth into a blender. From there you just need to dip the cuttings you are trying to root into the blended material before planting.

3. CHE

Ryan's Story

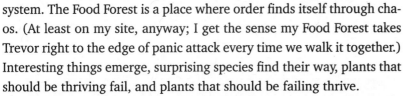

It's May of 2023. I'm strolling through my forest garden, coffee in hand, and taking notice of everything waking up. This time of year, I like to take stock of the flowers and fruit that have made it through the late frosts. A signal of how much canning and juicing I'll get to do at the end of the season. This is also a time for me to celebrate what works and take note of what may be struggling in the system. The Food Forest is a place where order finds itself through chaos. (At least on my site, anyway; I get the sense my Food Forest takes Trevor right to the edge of panic attack every time we walk it together.) Interesting things emerge, surprising species find their way, plants that should be thriving fail, and plants that should be failing thrive.

I'm deep in thought when I notice the Che tree and my heart sinks. The Che tree, also known as Chinese melonberry or the Oriental strawberry, is nine years old and just starting to come into maturity. Last season it pumped out an enormous amount of fruit, but this season the tree has failed to leaf out. A quick scratch on the bark, and sure enough, the tasty tree is dead. I peek closer to the base of the tree and notice some shoots below the graft line sprouting out.

We had a tough spring following an incredibly warm winter. The sap in all my trees rose almost a full month early this year, then in early May, there came a 17-degree night. This tree, or rather the graft, got toasted, a victim of the weather weirding that happens in the Valley. I'm always upset when I lose a tree. On the one hand, botanical casualties are part of the risk when committed to building a dynamic ecosystem that requires very little input from me. On the other hand, this tree is different. This tree is my son's, and it's become one specific plant I feel deeply connected to. The following is the story of how the tree came to this property and my son came into this world.

There I was, in the middle of CSA harvest when I got the call. My wife Joy's water had broken, and it was time to head to the hospital.

After having our first child in the hospital, we dreamed of doing the hippy thing and having an at-home water birth. Good judgment won out that day, as we did the math on how long it would take to get to a hospital from our rural mountain cabin should there be complications. This turned out to be a good decision. Joy's labor was long and hard. She worked for well over 24 hours, and the birth came down to the wire in order to avoid a C-section. The stakes felt raised considerably when Joy started pushing and the baby's heartbeat disappeared. The nurse looked at the midwife with concern. The midwife shrugged it off and stated, " It's time to get baby out." She turned to me and asked if I wanted to catch the baby.

The conversation went something like this.

"Do I have to wear gloves?"

"Your hands are clean aren't they?"

"Yes."

"Then get over here."

I got into position to catch the baby, with the midwife behind me. By this time, a pile of hospital staff had gathered in the room to watch a husband catch the baby, something rare at the hospital we were at. Joy did her heroic thing. I can remember staring at her pushing, a vein popping out on her temple, sweat covering her body, and a guttural sound releasing from somewhere deep inside her. This spell was broken by the midwife yelling at me.

"Are you going to let that baby drop on the floor? Wake up, get in there!"

And she kept yelling at me.

"Stop being so scared, grab that baby and hold on tight, you won't hurt it."

As I grabbed onto the slippery head and pulled, reaching my right hand under the body, I realized why the heartbeat had stopped. In a swift movement as if she was planning it all along, the midwife reached around me and slipped the umbilical cord from around our son's neck.

"Give the baby to momma."

I put our son on Joy's chest. Our baby was blue and still. The room went quiet. I know it's a cliché, but I need to say it anyway: ten seconds felt like ten minutes. I could hear the heart beating in my chest, the whole room was in shock. People started looking at each other nervously. The silence was cut by a baby letting out a large wail. The room relaxed.

That's when the midwife spoke up. "It's time for the placenta."

"I have to get the cooler." I responded.

I'll never forget the look on the midwife's face when I asked my father if he had the cooler. She gave me a look—that look kindergarten teachers give their students when they're being a little too squirmy at naptime. At this point, we had already been through a lot.

"What's that for?" she asked.

"We need his placenta," I stated matter of factly.

We had already fought with the hospital to allow us to take the placenta home. The midwife just shook her head.

A half hour later, my father walked out of the hospital with one of those 1980s beer coolers packed to the brim with ice and Joy and our son's placenta. He got to our house and into the freezer it went. Like a bag of summer squash or that uneaten ice-cream cake someone brought to your house and you felt bad about throwing out, the placenta got forgotten in the back of the freezer.

A year later, while cleaning out the freezer in order to get ready for the new season's bounty, we came across a bag full of frozen bloody mass. Our earlier plans to eat the placenta had turned on us. Placentophagy was now an interesting academic anecdote, no longer a practical bucket list experience. We would not be making lasagna out of this.

We didn't want to compost it either. That felt wasteful, and as our son Tamayo reached his one-year birthday, we were looking for a small ritual to celebrate his life. The year before, I had been up at Edible Landscaping with the fall Shenandoah Permaculture Institute course and found myself gobbling up ripe Che fruits off one of their trees. I shared this memory, and that's when we decided it was time to plant Tamayo's placenta with a Che tree. The memories have been connected ever since, and Tamayo proudly calls this tree on our property his tree.

This brings me back to the problem at hand. My son's tree is dead. I'm not looking forward to sharing this with him. As sad as I am, he will be heartbroken. To top it off, our annual ritual of gathering around the Che tree to fill up on sweet red fruits and tell Tamayo's birth story won't happen this year.

But …

There is hope in life coming from the rootstock. Pushing out of the base of the tree is a thorny Osage Orange looking to become a tree of its own. I think for his tenth birthday, we will hold another and altogether different ritual. This is when he'll learn how to field graft and take responsibility for nurturing Che fruit back into the life of the tree just like his placenta did a decade ago.

How to Grow

The Che tree (*Maclura tricuspidata*) is also known as Chinese melonberry and is native to central and east China. The tree is a relative to mulberry and fig and thrives in zones 6 to 9. Most people who eat the fruit describe the taste as a combination of fig and watermelon. We think it's sweeter than that. Every September our tree is loaded with beautiful seedless red fruit that tastes like watermelon with powdered sugar on top. It's by far the sweetest fruit we have in our system, and the kids love it. Many a September evening, our family can be found surrounding the tree and eating our fill until all head back to the house with sugar belly.

Integrating into Your Plant Guilds

This tree does well in full sun, so choose a location that receives at least seven hours of direct sunlight each day. Ensure the soil is well-drained. If your soil is mostly clay, consider amending it with organic matter to improve drainage. Like many fruit-producing trees, pruning is essential for maintaining the health and productivity of the Che tree. In late winter or early spring, before new growth emerges, prune away any dead, damaged, or diseased branches. This helps improve airflow, reduces the risk of diseases, and encourages new growth.

The Che tree makes a great anchor plant in the orchard, or can be a unique standalone landscaping plant, especially in children's gardens.

We have also put Che in our chicken yard to provide a late-season food source for our chickens.

Cultivar Recommendations and Propagation

We get our Seedless Che's from our friends at Edible Landscaping in Afton, Virginia, who graft them onto Osage Orange rootstock to prevent suckering. The story of how Edible Landscaping discovered the Seedless Che that they still propagate and sell to this day is a great tale unto itself. According to Michael McConkey, founder of Edible Landscaping, it used to be that the only Ches that were available in the nursery trade required a male pollinator and developed large cotton-like seeds that made the fruit hard to eat. One year in the late summer, one of Michael's employees found an old Che tree in Ivy, Virginia, that happened to be ripe, and when he sampled it, he found that it was completely seedless. He told Michael and they went out and realized there were no other Che trees around and concluded that it must be a self-fertile variety. Realizing they had found something unique, they took cuttings and were able to get them to root and propagate. The Osage Orange grafting technique had been taught to Michael by a professor named Darrow and was used to create a plant with less suckering than the Che on its own roots. The Seedless Che on Osage Orange rootstock is still available at Edible Landscaping, but if you are interested, try and reserve one early because they sell out fast!

4. AMERICAN PERSIMMON

Ryan's Story

(Adapted from the essay "Painted Leaves" previously published in *Rock & Sling* Fall 2019 Issue.)

I am 90 minutes into the stalk when I notice the wind shift direction. I feel it pressing against my neck. The sweat under my hoodie starts to feel cold, and I know I smell strong. I panic for a brief moment then relax. The bow is already drawn, I'm sighted in, now wait.

Be patient.

I spotted him downhill bedded with his back to me while walking to my stand at daybreak. I noticed the wind was in my face, so I took off my shoes, to make the stalk quietly, dropped to the ground and began a slow crawl. I needed to get within 20 yards for an ethical shot, and I had 80 yards to crawl without being heard or smelled.

Crawling 80 yards in over an hour is like chewing a single grape for 10 minutes. It's a stupid feat of mindfulness, an exercise in focused restraint. Once in position, I took another 20 minutes to stand, position my feet, knock an arrow, and draw.

My heart is beating out of my chest, and my body is freaking out. I have to hold my shit together. And that wind—I'm going to get busted.

He smells me and stands. The next few seconds are as long as the last 90 minutes. We make eye contact, he's disoriented, perhaps in disbelief I've gotten so close, which buys me another half-second. I release the arrow and immediately sit down, to hide my profile. I don't want to spook him hard resulting in a long tracking job. I hear the thump of the arrow making contact, I see the hind legs kick—a good sign. He bounds off and then stops 10 yards from where I shot him. He looks back in my direction but can't see me. I can see the blood spilling from his wound, painting the leaves of the forest floor red. He falls to the ground.

I weep.

I can no longer control the movement in my body. Everything shakes. A flood of relief and sadness and gratitude and joy overwhelm me.

It's a Blakean grain of sand moment. We only get a few of those in a lifetime, so I drink it in. And I sit in the sun, waiting. Thinking.

Suddenly, I'm cold.

This is not the first meat I've harvested from the forest. I've accepted that, as a meat eater, blood must be shed. There is so much more to this release—it's hard to unpack—and, honestly, distracting.

Like a Zen koan, the effort to separate the pieces doubles in on itself only to explode into a million pieces of disconnected logic. I don't try to put them together.

This always happens.

The last decade was hard. A few years ago, the rain came and the farm business failed. It didn't happen fast. The scene unfolded slowly. By July, we had a full creek crossing our driveway into the lower fields that didn't stop flowing for two years. I limped out of the season—organic vegetable farmers don't get subsidies or insurance coverage like the big corn, wheat, and soy guys. Before then, we found ourselves front and center in a public battle over a controversial pipeline. The local limelight shined bright on our family. We just wanted to keep farming. Instead, we became the fulcrum of a political lightning rod that played itself out like a microcosm of the tribal politics we currently live in. Allies and foes traded places, everyone had an opinion, and generally speaking, our community lost its shit.

And before that, my friend killed herself. As a trained counselor, I've dealt with suicide before, but not like this. I'm left with questions of what I didn't do, where I failed her family. I essentially took out a mortgage on my brain to earn the training to become a mental health clinician. And I couldn't even help a friend.

So I farm.

I farm hard, family time is rare in the spring and summer.

At a certain point, it starts to feel like a pile on.

Decisions have consequences. I have chosen my path on the land.

Of all the upside that this life connected to the land has, there are dark, ugly, and maddening parts. A cliché, but true: intimacy is beauty and love and sweet sweat, but it's also blood and shit and disappointment.

Things don't always go right—this too is what we have asked for. To homestead and farm, to tinker, grow things, create … is to fail.

Chores still need to be done.

Before tragedy and failure, chop wood and carry water; after tragedy and failure, chop wood and carry water.

I've gotten ahead of myself.

I notice during the wait an American persimmon tree. Everybody swears by their fruit. It seems there's always some baby-faced hipster trying to convince me why persimmons are the best fruit.

Thirty minutes pass.

It is now time to dress the deer. I run my blade up the belly and split the ribcage. I plunge my hands in. It's warm and smells of sweet earth and metal. The relief of the warm blood on my cold hands cause me to pause for a minute, elbow deep in the animal. I then pull the guts out, carefully separating the kidney, heart, liver, and caul fat to be placed in a ziplock freezer bag. I leave the guts in a pile for the coyotes to feast on tonight. But not before noticing a pile of sweet sticky persimmons oozing out. There it is again, the persimmon. I wonder how many persimmon trees have been planted after passing through the guts of a deer.

The forest eats everything, even the bones. Nothing is wasted. Not even a cliché passed off as introspective profundity. From this perspective, all things matter and nothing matters. Lao Tzu was an asshole.

The animal is too big to throw over my shoulder, so I wrap my hand around an antler tine. The feeling conjures memories of hours spent transforming raw palm flesh into callous. In younger days, that sensation came from dunking basketballs or lifting weights. These days, it's an axe handle or a garden hoe. Gripping an antler is somehow different. More organic. The tines of an antler have that same charisma of a perfect walking stick that's spent years being choked by honeysuckle vine.

I grab a handful of persimmons off the forest floor before the long, bruising drag out of the woods.

On the drag, I think about the persimmon. The sticky sweetness mixed with the sour ferment of the animal guts. The overripe ones I've just tucked into my hunting pack. I've never wanted to eat one this bad in my life. The act of killing something is intimate and horrifying. Even for sustenance, it is no small thing to take a life. The experience leaves me childlike—open to impressions—and right now I'm imprinting hard on this weird little fruit.

I've got the animal back at my farm, and he is hanging in the walk-in. I'll butcher after he's hung in 37-degree temps for a couple of days. But first, I need the tenderloin for dinner tonight. I love how the meat pulls away from the bone just by running my hands up against the back of the ribs. It's like running your hand up the side of the wall looking for a light switch in a dark closet. The revelation comes as the meat pulls away into a perfect cut.

In the kitchen now. I cut the meat into medallions. The knife glides through with all the ease and pleasure that comes with a perfect fit. It seems to say, a sharp blade was meant to be here. A rub of black pepper, juniper berries, cayenne, salt, and garlic powder. And I place the dish in the fridge in order to go clean up.

My family sits around the table tonight. They protect me from the sting of reality, from my mistakes, from the cold feedback that sweeps into a life. I don't want it to end. I want to stretch it out. Have it linger like a big heavy wine on the tongue long after being swallowed.

We eat persimmons with it.

Off-farm work, school, sports, side hustles, and projects leave our family stretched. But tonight we are together. Every movement I've made from first light until now has been about this Moment.

Here and now.

How to Grow

American persimmon (*Diospyros virginiana*) is one of the most widely adapted, easy-to-grow fruit trees in the temperate world, making it a quintessential Permaculture plant. It can grow in a variety of soils including degraded clay, is drought tolerant, and is virtually pest and disease resistant. The one limitation for northern growers is that most varieties only grow to Zone 5, with a few hardy to Zone 4.

Oh, and did we mention that it is an absolutely delicious and underappreciated fruit with a variety of uses for humans and animals? More on that later.

Integrating into Your Plant Guilds

Wild, unselected persimmon trees can grow up to 60 feet tall. However, the 90-chromosome cultivars we recommend (see below) grow to about 20 feet, making them a perfect tree to integrate into a diverse linear guild Food Forest or U-Pick system. These varieties are self-fertile and produce mostly seedless fruit, which means you don't necessarily need to plant a male. Like any fruit tree, plant at least two in the same area for good pollination and production.

Persimmons are also a premier crop to grow as livestock feed and have a long history of use in silvopasture systems in the United States, especially as a food source for pigs.[2] The persimmon fruit is high in calories and nutrition and drops throughout the fall and into the winter, providing a food source during a time of year when forage is scarce. On a small scale, consider planting some persimmons around the chicken or hog yard. On a larger scale, you can design an entire tree crop-based hog production system by integrating persimmon with other silvopasture trees including mulberry, chestnut, oaks, and hickories.

Cultivar Recommendations

Native persimmons come in basically two kinds—a 60-chromosome type that is larger and native to the southeast United States, and a 90-chromosome type that is smaller and native to the Midwest, central, and southern United States. Because of its smaller stature and generally larger fruit, most named cultivars available to growers are of the 90-chromosome type.

One of the first and most well-known cultivars is the Early Golden, which is a parent of many of the other cultivars that have since been developed. A great source for named cultivars of persimmon, especially cold-hardy varieties, is the nursery Perfect Circle Farm in Vermont. Owner Buzz Ferver says persimmons are his number one favorite fruit, and he has collected genetics from all over the country, many of which would be lost to history without his preservation efforts. Buzz's favorite tasting varieties include the hybrid Nikitia's Gift and the American cultivars Mackenzie Corner, H63A, Geneva Red, King Crimson, and Juhl.

By the way, there is a whole world of Asian persimmons out there if you are really into this tree. These varieties, which have a long historical use in East Asia, are smaller and less cold hardy, but are a great addition to a Food Forest in the warmer climates of Zone 7 and above. They include some delicious varieties that can be eaten firm off the tree and have less tannic properties than the American persimmon.

Propagation

American persimmons can be grown from seed after a period of cold stratification, but to get named cultivars, you will have to purchase grafted trees or graft your own. One growing challenge with persimmons is that since they have a deep taproot, like pawpaws, they can be a little difficult to transplant or at least slow to establish. When transplanting, try to select a young tree in a deep tree pot that is not too rootbound, or purchase bare-root first-year seedlings from trusted nurseries that produce robust root systems.

Harvests and Uses

The Latin name of the genus, *Diospyrus*, translates to "Food of the Gods" and indeed a truly ripe persimmon is divine. Beware that if the American persimmon is not fully ripe and soft or even mushy in texture, it will be very astringent and will make your mouth pucker when you bite into it. At its peak though, the fruit tastes like an apricot mixed with a peach with hints of cinnamon and caramel.

While it is a delicious fruit to eat out of hand, many of the traditional uses involve removing the persimmon pulp for use in other delicacies. The annual Mitchell Persimmon Festival in Mitchell, Indiana, is famous for its contest for the best persimmon pudding, a classic Midwestern dessert. Persimmon is also a traditional base for beer and other alcoholic ferments due to its high sugar content and complex flavors. For the hunters out there, persimmon wood is a favorite material for crafting bows.

5. PAWPAW

Trevor's Story

I didn't start consciously building relationships with plants until I was in my late teens, but looking back, they were always there, waving to me, inviting me, keeping me shaded and cool. Growing up in downtown Richmond, I spent my childhood exploring the rapids and trails of the James River Park System. My best friend Will and I would ride bikes down to Belle Isle and swim and ramble all day until the six o'clock dinner bell summoned us home. Later on I got really into mountain biking and trail running. Some of my best memories from my early teens were riding the single track-trails along the river with my Dad.

It wasn't until I was about 18 that I realized I had been growing up under the shade of what would become one of my best plant friends, the pawpaw. This native tree literally blankets the understory of the forest across the entire James River Park System, as well as most river systems in the Mid-Atlantic and Ohio River Valley.

When I discovered the pawpaw tree in college, I became obsessed with finding and foraging its delicious fruits. My sister, a few years younger than me at the same school, never let me forget the time her friends spotted "that weird guy climbing in the bushes on campus" only to realize later it was me! One time I rolled up to a party with home-made pawpaw pudding bread and started handing it out to people and enthusiastically explaining, "This is North America's largest native edible fruit!" This turned out not to be the conversation starter I hoped it would be, although the naturally sweet spongy bread was devoured by the half-drunk party goers within minutes.

Now, of course, I wasn't the only one who knew about the pawpaw, but it sure seemed like it had disappeared from mainstream consciousness amongst the city folk I grew up with. The only vestige of its cultural legacy in my early life was that little ditty "Way Down Yonder in the Pawpaw Patch" that I learned from my mom who surely learned it from her mother, Skip, who herself grew up in the epicenter of the pawpaw's natural growing range, the Ohio River Valley. I'd love to know, did my grandma's parents have their own stories about the pawpaw, their own relationships?

I am remembering now, I was with my mom the first time I ate a ripe pawpaw in the wild. I was home from college for fall break, and my Mom and I were hiking at Texas Beach, one of our favorite places along the James River. We had waded out to a small island to get away from the crowds and were bushwhacking across it when I spotted the greenish-gold orbs dangling from a small tree. "Mom, I think this is a pawpaw! Should we try it?!" My Mom was always down for a new adventure. "Sure," she said. I cracked it open and scooped out some flesh for both of us. Soft, sweet, and only slightly bitter around the seeds. "Tastes like mango-banana pudding," she declared.

We lost my mom to cancer in 2023. She was only 71 years old and had just welcomed her fourth grandchild into the world. One of the many layers of the heavy grief that still hangs over me is that I won't get to go on any more adventures with her. She was a theater artist with a much different background than me, but she was always in tune with the natural world. So when I discovered my love for plants, she not only supported it but jumped into the fun. Now that I am a parent, I can imagine how cool it must have been for her to watch me discover my passion and then learn with me as I chased it down.

Towards the end, her illness had altered her personality—she didn't speak much, and when she did, it tended to be on the surface, not the deep conversation I craved. Every once in a while, though, her true self would shine through. The last time she visited our farm, we took a long, quiet walk together. I held her arm up and down the hills, and we walked through the berry patch to the creek and then down along through the young pawpaw patch and finally back up to the walnut grove. My Mom stopped under the shade of a large walnut and started slowly sweeping an outstretched arm across the landscape in front of us. "Trevor, this place is so beautiful," she said softly. "You must be so happy here. You must know every inch of this place like the back of your hand."

How to Grow

We both have small plantings of young pawpaws (*Asimina triloba*) on our farms, but we are by no means experts on growing them yet, so we caught up with our colleague and fellow fruit-growing enthusiast Adam Dusen to learn more about pawpaw production. Adam owns Hundred Fruit Farm in New Hope, Pennsylvania, and he says pawpaws are the most popular fruits that he sells there.

"One thing you hear that is a little bit of a myth is that pawpaws love wet soil," Adam says. "They are more tolerant of wet soil than other fruit trees, but they will be happier and grow much faster in well-drained soil." In addition, Adam says that even though they can grow in the shade, which is their natural habitat, he recommends growing them in full sun like other orchard trees for maximum fruit size and production. In the southern U.S. and other hot climates, they might benefit from some partial shade when they are young and getting established, but in the Mid-Atlantic and North they do better in full sun.

Pawpaws are almost completely pest free, and Adam has never sprayed anything, organic or otherwise, on his trees, and they are thriving. What's more, he says they are the most deer-resistant fruit tree he grows on his farm. Although it is important to note that young pawpaws, like all young trees, can be severely damaged or killed by deer rub. Another beneficial trait of pawpaws is that the trees can re-bloom after a late frost event, allowing for a decent fruit set even in difficult growing seasons.

In terms of pruning, Adam has a simple and easy approach where he lets the trees grow to about eight feet and then cuts back the central leader, which stops the tree from growing vertically and keeps it at a manageable height for hand harvesting.

Integrating into Your Plant Guilds

One challenge with integrating pawpaws into a guild layout is that they really need to be planted close to other pawpaws—at least within 15 feet—to ensure pollination. Adam recommends planting a minimum of three together in a patch and adding diversity in the understory with ground covers like lemon balm, mint, and dwarf comfrey. In a mature stand of pawpaws, Adam says you could experiment with shade-loving native ground covers such as wild ginger and dwarf crested iris.

Cultivar Recommendations

"The named cultivars are so much better than the wild ones in my opinion. It's really, really worth growing the grafted ones," Adam says. Wild pawpaws are smaller and more seedy while the cultivars are large and have smaller seeds. In particular, anything in the Peterson series, developed by pawpaw breeder Neal Peterson, are excellent, as are the selections out of Kentucky State University. Standouts include Susquehanna, Shenandoah, KSU Benson, KSU Chappell, and KSU Atwood. Good non-patented cultivars include Sunflower, Overleese, Avatar, Freebyrd, Sri Gold, and the late Jerry Lehman's selections Jerry's Big Girl and Maria's Joy.

Propagation

"If you're a DIY person, you can learn to graft, and eat your pawpaws and plant the seeds then graft onto those seedlings, and you get pawpaws for really cheap," Adam recommends. He emphasizes that pawpaws really grow well with an undamaged taproot and do not transplant well, so you want to buy young plants grown in tree pots or direct seed into the ground in their permanent locations.

Harvest and Uses

In our foraging days where we were working with very tall understory trees, we would typically shake the trees and pick them up off the ground. In Adam's pawpaw patch, the trees are small enough that he can harvest them by hand right off the tree by snipping them off with a pair of pruners, which prevents the fruit from dropping and getting damaged. The key is to avoid bruising that will speed up ripening and shorten their already short storage life.

In addition, Adam harvests the fruits when they are just starting to ripen and puts them in the refrigerator where they can keep for up to two months. When he is ready to sell or eat them, he just takes them out and lets them ripen at room temperature, similar to an avocado or peach.

Our favorite way to eat pawpaws is fresh. Cut them in half and scoop the custard-like flesh out directly with a spoon. But there are also endless possibilities for value-added products and recipes including ice creams, beers, breads, and pies. During pawpaw season, a single pawpaw makes a nice standalone light breakfast, boasting about 80 calories, one gram of protein, and all of the essential amino acids. Be careful not to overconsume pawpaws, though, as they do contain a compound called acetogenins, which are thought to have potential neurotoxicity if eaten in gross excess.[3] (Since there is not a long history of pawpaw consumption in modern times, more research needs to be done on this.)

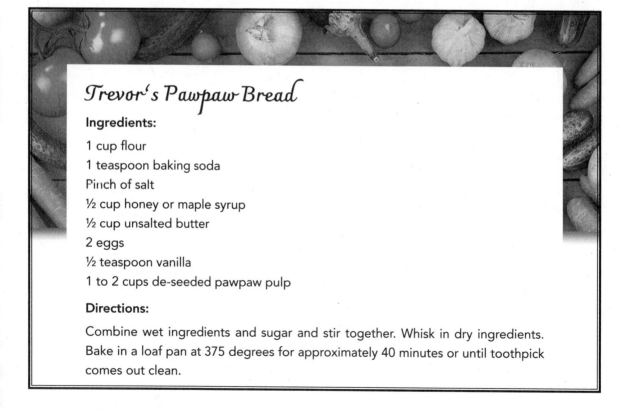

Trevor's Pawpaw Bread

Ingredients:

1 cup flour
1 teaspoon baking soda
Pinch of salt
½ cup honey or maple syrup
½ cup unsalted butter
2 eggs
½ teaspoon vanilla
1 to 2 cups de-seeded pawpaw pulp

Directions:

Combine wet ingredients and sugar and stir together. Whisk in dry ingredients. Bake in a loaf pan at 375 degrees for approximately 40 minutes or until toothpick comes out clean.

Budding willow,
CREDIT: CELIA RUTT

Center left: *Pile of red clover,* CREDIT: CELIA RUTT

Center right: *Elderberries,* CREDIT: CELIA RUTT

Left: *Crowning rhubarb,* CREDIT: CELIA RUTT

Top: Emilie Tweardy, Ryan Blosser, Trevor Piersol, cofounders, SPI, CREDIT: NICK FAIRCLOTH

Left: Willow flower, CREDIT: CELIA RUTT

Right: Pawpaw flower, CREDIT: CELIA RUTT

Left: *Zinnia*,
CREDIT: CELIA RUTT

Bottom left: *Trevor eating Che fruit*,
CREDIT: JAMAICA GAYLE

Bottom right: *Ryan with a 'Nikita's Gift' persimmon from his orchard*,
CREDIT: RYAN BLOSSER

Top: *Ryan in his favorite place, in front of a class,* CREDIT: JAMAICA GAYLE

Bottom: *Emilie and Ryan in discussion during a PDC,* CREDIT: JAMAICA GAYLE

Top left: *Che fruit,* CREDIT: JAMAICA GAYLE

Top right: *Jerusalem artichoke,* CREDIT: JAMAICA GAYLE

Center right: *Stinging nettles,* CREDIT: CELIA RUTT

Above: *Rhubarb stalks,* CREDIT: CELIA RUTT

Left: *Flowering rhubarb,* CREDIT: CELIA RUTT

Above: *Black locust flower,*
CREDIT: TREVOR PIERSOL

Top right: *Honeyberry,*
CREDIT: CELIA RUTT

Above left: *Honeyberry in hand,* CREDIT: TREVOR PIERSOL

Above right: *Indigo bush,* CREDIT: RYAN BLOSSER

Left: *Comfrey flower,* CREDIT: CELIA RUTT

Top left: *Strawberry*, CREDIT: CELIA RUTT

Top right: *Ryan and Trevor thinning peaches at Wild Rose Orchard,* CREDIT: CELIA RUTT

Left: *Elderflower*, CREDIT: CELIA RUTT

Below: *Wild Rose Orchard linear guild Food Forest*, CREDIT: SCOTT TURNER

Above: *Daylily,* Credit: KaiMarley Blosser

Center right: *Blooming redbud,* Credit: Celia Rutt

Bottom left: *Piersol family in the orchard,* Credit: Trevor Piersol

Bottom right: *Yarrow,* Credit: Celia Rutt

CHAPTER 4

Nitrogen-Fixers

1. REDBUD

Ryan's Story

THE HEARSE IN THE DRIVEWAY is staring me down. Ice in my glass rattles. Raindrops, the thick kind, roll down the window. I notice the redbuds are in bloom. Not so much red, but a purple pop among a gothic landscape of thin lines and gray sky. I sip and remember.

Warmth slides down my throat, burning, down to the stomach. I shiver when the cold emotion and whiskey mix. I remember this feeling fondly, maybe it's the whiff of a thrill that seems to coat the air of melancholy. The casket is open in the living room, but I won't look, the redbuds have my attention instead, and I remember …

"Hey boy. Get up. It's already five o'clock, we gotta get over at Windmill Market."

Me and Dad, waiting at a country convenience store parking lot littered with cigarette butts and Styrofoam coffee cups.

Don was a plump muscular man, with dirty teeth, a dark beard, and a bright smile. The kind of man who slapped you on the back when he said hello. Dad always said he had a horrible sense of direction, especially on fishing trips. If Don said we should go north, everyone knew south would be better. Don would take off north anyway, and the group would follow, knowing, but not caring, that it was the wrong direction.

"Where tha' hell you boys been?" he asked, snatching a couple of beers and a Yoo-hoo from the boat cooler attached to his blue Jeep.

"You're late," Dad said.

"D'you wanna' catch some fish or not?" asked Don, grinning and holding up a shiny new bottle of whiskey.

Dad nodded and we climbed into the truck to begin the drive up old Calf Mountain Road towards the lake. The men shared the bottle, while behind us, a green Jon boat rattled against the trailer.

The March sun was peeking over Calf Mountain, warming the lake and the surrounding forest. Everything was still dressed in police blue from the night before, and the air felt crisp, cold even.

Don fired up the motor, and the boat leaped forward, cutting through fog and water. The morning was lost to this, racing from one end of the lake to the other.

"Son, grab us a couple of beers—and grab lunch while you're in there."

I handed a beer to Don, shaking father's first.

"What?"

Laughter and sticky bubbly drops cleansed my Dad. He sat grinning, licking his hands.

"Stop messin 'round you little piss ant," Don said. "Hand me that food."

What made these two men, Don and my Dad, click? Their friendship was a perfect circle. Dad, he was quiet, wise, good. Don was loud, crazy, free. Dad bathed in this freedom and made Don good. They loved each other for being what the other was not.

Don sitting in the back of the boat, shoulders held taut, jerked his rod hard above his head, knocking pipe and hat into the water.

"Shit. I missed 'em."

"You're missing something else." Dad got a strike before he could finish the thought. This time he landed the fish, and it was on. We found ourselves in the glory of the trip. Catching, not fishing; floating under a patch of blooming redbud trees watching swills dot the surface around us.

Crack! The sky opened and nobody noticed.

Crack! And a flash of lightning.

But the fish were biting.

At this moment, there was a light tug on my rod. Something was nibbling very lightly. Not knowing what to do, I began to reel the bait in very slowly when something hit the end of my line like a dog latching onto the neck of a loose chicken. My rod doubled over. Don and Dad turned from their posts to coach.

"Take it easy son."

"No need to panic."

"Let him tire himself out. Just hold it steady."

The line moved towards the boat.

"Now reel!"

I reeled as fast as I could. I didn't feel the fish on the line, but the men kept yelling so I kept reeling. Suddenly, right in front of the boat, the fish jumped and then dove.

"Shit. D'you see that? Must be five pounds!" Don shouted

"Don't let him under the boat," Dad kept coaching.

I held tight.

"Now, pull the tip of your rod up slowly, don't reel, I might be able to reach 'em."

Dad netted the fish, and I slid down in the boat, exhausted. The men toasted the catch with a taste of the whiskey.

They smelled blood.

While I stood over the bucket staring at my trophy, the men fished furiously. Then, the crack in the sky got a whole lot bigger, and hail dumped from the clouds.

I started crying.

"Calm down, the storm'll pass."

"But lightning," I said.

"We're fine," said Dad turning to Don. "But you know, we probably should head back, it's getting dark and the ramp is an hour away."

"Come on man, we've been in worse shit than this."

"Yeah, but not with the kid." Dad looked towards me.

"Ahhh, it's good for him. Here, have a drink."

Dad stared hard at Don.

"Alright, give me a sec."

We pulled out of the cove, the Jon boat chugging along, and pushed into the middle of the lake, thinking that straight across was the best way to get to the ramp. As we moved further away from land, the little Jon boat rocked. Blowing and howling, the wind scared up whitecaps across the landscape of the lake. Peaks and troughs smacked the boat one after another.

"We can't make it this way," Dad announced.

"I'm gonna to take her over there so we can hug the bank the rest of the way."

"Can't we hurry?"

"Relax son."

Lightning struck the bank next to us. A tree caught fire.

"Hey can't this thing go faster?" Dad yelled.

"I want to go home!" I screamed.

"Stop crying."

"But I'm scared."

Don sat quietly brooding over the fastest path back to the ramp while my father tried to calm me.

Lake, ground, sky shook violently as greens and browns of lake and trees darkened.

"Ahhhh!" I screamed.

"I ain't playin' boy, shut up. You've got to shut up!" Dad snapped.

I sat down and cried quietly. In the next moment, there was a break in the clouds.

"We're gonna make it boys!" shouted Don.

"You love this shit don't you?" Dad asked.

"Yeah, boy." Don was standing up in the back with his hand on the motor, maneuvering the shallows as rainwater dripped down his face. "Isn't this beaut?" A thump on the bottom of the boat interrupted his flow. "What the …?"

Dad said. "You ran the motor into the bank."

"Help me get this thing into the boat," Don shouted. " One, two, three …"

"Oww!" Don held up his hand. The tip of his finger was missing, and from it blood was pouring down his arm over his shirt and into the water. The motor dropped and crashed onto the back of the boat. The back sank into the lake water filling it up.

"Uh oh."

The men pushed the motor back over the edge of the boat. Don took off his shirt and wrapped his hand. Rain and blood was dripping, racing down, slaloming around the hair on his round belly.

"We gotta move!" Don shouted. "Son, get back here and start dishing this water out."

Dad and Don paddled in rhythmic grunts. We were moving faster, and I was distracted from the storm for the moment, knowing only the grit of the bloody lake water.

Chaos and threat softened into safety as the boat reached the shore just as the storm passed. In the parking lot, exhausted and hungry, the men left me for a moment to hunt down snacks at the convenience store.

They shuffled out of the store, Don with his head hanging and clutching his hand to his chest, Dad smoking a cigarette, shoulders bent. They stopped short and Dad looked up. He tapped Don on the shoulder and pointed.

There I was, standing in front of the pickup after salvaging what I could from the back of the boat, eight years old, fighting back tears and shivering from the cold. My shirt was torn, I could feel the bruises setting on my face and hands, my arms were smudged with dirt, and I was covered in Don's blood. I lifted up the bottle of whiskey like it was a prize fish for the men to see.

Dad lifted his shoulders, and Don raised his head, a broad smile radiated from their faces. I was miserable and felt happy and started to hate myself for wanting so badly to be home an hour ago. They walked over to me, Dad grabbed my hand and led me to another redbud tree where we sat down.

"Hand me that bottle, boy." Don reached for it and took off the cap. "Ahhhhh." He handed the bottle to Dad.

Dad took a swig and swung his head around.

"Here boy. It'll warm ya' up."

I took the bottle and drank deep. Burning and coughing and gagging. Then, just as my eyes were watering, a warm sensation starting deep in my chest spread throughout my body. I felt good and stretched my legs out to sleep. As I drifted off, I heard laughter and conversation coming from the men, and I thought of that redbud, the whiskey, and blood.

"Hey Ryan. Come here, I've got something to show you."

My head snapped away from the window to see Dad, dressed in a gray beard now, eyes sunken in sadness and satisfaction, his face making a smile. I wasn't sure how to comfort him, Don and I were close, but he and Dad were like brothers. The cancer that ate up his liver was a stinging reminder of a hard-lived life, one that made Dad flinch.

"What's up, Dad?"

"Look at this."

He handed me an old beat-up bottle with a label missing. It was caked in dried blood and dirt. The bottle was half empty. I fought tears. Dad laughed.

"He looked pretty bad, didn't he?"

"Yeah, it was bad."

I held up the bottle. "So this killed him?"

Dad grabbed the bottle from me and took off the top. He poured us both a glass.

"To Don?"

"To Don."

The liquor felt good going down.

Dad looked out at the yard, back at me and smiled. "Redbuds are blooming, it's time to go fishing."

How to Grow

Redbud (*Cercis canadensis*) is not the easiest native tree to get established, but it is well worth the effort for its multifunctionality and beauty. Native to eastern North America, it is a small to medium-sized tree that can grow anywhere from Zone 4 to 8 in a variety of soil types. Although it can eventually grow to 30 feet, it grows slowly and is more commonly about 15 feet tall and wide. Unless pruned for a single trunk, it will naturally form a branching vase-like shape as it grows.

Integrating into Your Plant Guilds

Redbud acts as a nitrogen-fixer, a beneficial attractor, and a human food source (its flowers are edible) in a guild planting. In a small orchard, it could actually be an anchor tree that is alternated with your other fruit trees. In a broadacre planting of large agroforestry trees, it is a great nitrogen-fixing support tree, like black locust, that can be interplanted amongst your primary crop trees.

Cultivar Recommendations

A handful of redbud varieties have been bred for the ornamental market, but we recommend sourcing yours locally from seed or even transplanting from wild patches near your garden. If you are really into the redbuds, there is a pretty cultivar called Alba that has white flowers.

Propagation

Redbud is best grown from seed in a tree pot or transplanted at a young age. The tree does not transplant very well, so we recommend growing your own from seed when possible and transplanting only once into their final location.

Harvest and Uses

Redbuds have a variety of uses in the ecosystem as a natural food source and habitat for insects and a nitrogen-fixer for depleted soils. They also have the added bonus of having edible flowers, which blossom in the very early spring before almost any other tree. The pink flowers, which are in the pea family, add a really nice pop to a spring salad. Our friend, the chef and urban permaculturalist Tom Parfitt, is known to make some of the most beautiful and artistic focaccias in the state of Virginia.

He uses redbud flowers liberally in his doughy art. Not having Tom's skill or creativity in the kitchen, we stick to fritters.

Redbud Fritters

Note: Black locust flowers can be substituted for redbud.

Ingredients:

⅓ cup flour of choice (we used rice flour)
⅓ cup cornmeal (we used blue)
1 teaspoon baking soda
Salt to taste or salt and savory herbs
1 egg
⅔ cup milk of choice (almond, cow, coconut)
1½ cups redbud or black locust flowers freshly harvested
Butter or olive oil

Directions:

Mix dry ingredients together (except flowers).

In a separate bowl, beat egg, and mix in milk.

Mix dry and wet ingredients. Dip and coat flowers in batter.

Melt butter or warm oil in pan on medium heat.

Cook until they start to brown. Flip and brown other side until crisp.

Drain on cloth. Enjoy.

2. FALSE INDIGO, AKA INDIGO BUSH

Ryan's Story

I began my farming career as a child and family therapist. Most of this work took place deep in rural America.

When we moved back home from Hawaii, I was trained as a poet—that's it—oh, and I wanted to farm. I needed a trade before I was going to be able to pursue my dream of being a first-generation farmer. The only thing I seemed good at was metaphors and sitting quietly while other people talked.

After graduating with my MA, and EdS, I became an in-home clinician. Often, my clients were court ordered and sometimes very deep in the holler.

There are dark things that happen in the mountains.

There are dark things that happen everywhere, and being an in-home counselor gives you a front-row seat to the nightmare and the struggle. I made mental health house calls to unstable people in fragile family systems living in economically depressed areas.

In one particular case, deep in one particular holler, I spent time with a remarkable family.

Their house was one of those that started as a double-wide and turned into something very different after years of shoestring budget add-ons.

I was assigned the case by the court system after a mother was sentenced to ten years of prison on a felony drug and money laundering charge. Back home, their father was left alone with his three kids, one of them just over a year old.

My first time meeting the family, I remember sitting on a couch having a conversation with the father. The room was chaos. Three children running from wall to wall. Nobody, including the father, had their shirt on, and the air conditioning was blasting in the tiny house, making it feel as cold as a beach restaurant. It had to be 50 degrees. The chaos stopped for a brief second when the four-year-old looked at me and asked;

"Want to see something cool?"

I didn't answer; instead, I just sat waiting.

That's when the little man ran up his father's legs and lap, up past his torso so he was standing on his father's shoulder. He then reached into his father's eye socket, popped it out, jumped off his dad, and scurried over to plop the fake eyeball into my lap. This all seemed to happen in one motion like a cartoon.

There I was, a master's-level clinical mental health counselor, sitting in a client's living room with a fake eyeball pulled fresh from my client's socket staring up at me from my lap and a little kid with crusty snot all over his face and three dead teeth in his front row laughing at me.

I handed the eye back to my client and fought hard the urge to go wash my hands.

He liked to be outside, the client. I had just started my plant journey and had an intense interest in anything outdoorsy. One way I earned his trust was through my own curiosity of the outdoors and his knowledge. We would take walks around his property, and he would identify plants. I noticed one week that there was a bush all over his property that had a strange growth habit for a shrub and the most haunting purple flowers that looked like their tips were on fire. When I asked about it, he told me what it was and just said one word.

"Turkeys."

"What?" I asked.

"They're for the turkeys, they love the stuff. All I gotta do is point my rifle out the window and shoot."

I wondered if this was the same rifle that his son pulled the trigger of while crawling across the floor leading to that fake eyeball.

We spent a great deal of time that summer outside. We worked, hard therapeutic work. Then something happened that changed everything.

One day, like every other day, I'm hanging out outside when a family member shows up. The client was just telling me a story about this family member and some of the struggles he was having. This family member got out of the truck and walked towards us with a purpose. He was smiling so the speed of his gait didn't seem alarming. When he arrived, he said he had something to show my client. He pulled out a pistol and started showing it off. This person was excited about purchasing a new firearm. That's when he rolled the pistol over in his hand and turned his body so that the barrel was focused at my head.

I sat still, like a bee had just landed on my nose, but this wasn't a bee. I let him speak.

"You and the government need to stay out of this area."

"Don't shoot him, he's not the guy," I heard the client say.

In my head, I wondered what that meant. Meanwhile, I didn't ask any questions.

The man lowered the weapon away from my head, walked back to his truck, and drove away.

Our session wasn't over so I stayed, but I doubt I was worth anything as a counselor.

For the next hour that we talked, I was a complete wreck. Maybe this was obvious to the client. Meanwhile, what stands out in my memory of the event is the plant next to me the entire time—the false indigo and those haunting purple flowers that looked like they were on fire.

Later I would discover that it is native, nitrogen-fixing, and incredibly functional.

Upon this discovery, I immediately ordered 20 of them and planted them in a native hedge on my property to remind me of that day up in the holler but also because:

"Turkeys."

How to Grow

False indigo or indigo bush (*Amorpha fruticosa*) is a medium-sized, deciduous, nitrogen-fixing shrub native to North America. In our area, it is an early pioneer species, which means it is resilient and easy to grow. Indigo bush can thrive in full sun or partial shade and is even tolerant of wet soils.

Integrating into Your Plant Guilds

Indigo bush is a great way to get nitrogen-fixers into your linear guilds without having to get clover established or planting large trees like black locusts. In our orchards, we have planted indigo bush every seventh or eighth tree. It has a tendency to spread by suckers, and for this reason, it is a great component of a privacy screen or wildlife hedge.

Propagation

Indigo bush can be started from seed or taken from cuttings. If you or a friend have an existing stand of it, you might consider digging a few plants up as root divisions in the fall and propagating them out to your guilds.

Harvest and Uses

The main uses of indigo bush are as a nitrogen-fixer and beneficial attractor, especially as habitat for larger wildlife like turkey, as Ryan's story can attest. Some of our students have told us they have made dye out of the flowers, which apparently contain small amounts of indigo pigment.

3. BLACK LOCUST

Ryan's Story

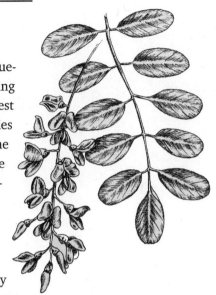

I bought a piece of land on a high of sheer enthusiasm and clueless energy in order to start a farm. I knew nothing. Everything was both possible and impossible. Aside from it being the rockiest piece of land in the entire county, there were only two tree species on it at the time: Eastern red cedar and black locust. On the one hand, I loved the fact that there were any trees of any size. The property was located smack in the middle of a multi-hundred-acre corn, wheat, and soy operation. On the other hand, these were scrubby ugly trees. The black locust—the ones with any size on them—were diseased and broken. I tend to appreciate the Tim Burton aesthetic in all things landscape design, but this was gothic to the max. I have a distinct memory of sitting on my deck, having just bought the house and locking eyes with a turkey buzzard sitting on a broken limb of a dying, leafless locust tree—and it was only about ten feet away! Washington Irving, eat your heart out.

Those first couple of years, I spent all the free time I had ripping the seedlings out of the ground before the thorny trees took hold. Eventually I learned that the flowers were edible and started harvesting them in the spring for fritters. This helped a little bit. Another thing that helped is that in all of the Permaculture books I was reading, the tree was listed as an important nitrogen-fixer. So that seemed cool. But my interest and relationship to the tree never fully formed until I met my neighbor.

This is the part of the story where, normally, I'd share with you the name of said neighbor. But dude is allergic to attention and to name him in a book would no doubt inspire a panic attack. Often, when I ask for permission to take his picture, he says, "I'm not prepared to be famous." So let's call him the Soul Valley Wizard or the Wiz for short.

I first met the Wiz at a local meeting of folks interested in living sustainably. Back then, the movement was referred to as green. After the meeting, this wiry old man with a fairylike grin and a mischievous sparkle in his eye approached me and asked, "How green are you?"

through a Puck-like smile. In the conversation that followed, we soon learned that we were neighbors. He invited me over the next weekend. Turns out, the Wiz was and is just about the most sustainable man I know!

My first visit to the Wiz's place was like entering an Ewok village. Recycled corn silos turned into buildings with catwalks high up in the trees, junk turned into art resembling robots or aliens integrated into the property. He retired from architecture in his 30s and has spent the last 40 years building up his property through collecting other people's junk. It was organized, workmanlike, and remarkably creative.

The space was also functional. Immediately, I knew I wanted to learn from this man. I have learned many things from him. One of those is a practical reason to love the black locust.

The Wiz, like me, had a little Jøtul wood-burning stove. Like me, his only heat source was this stove. The difference? I ordered pre-cut wood to burn in my stove, paying someone else to harvest the tree and break it down into pieces for me. Wiz, in his 70s and still moving around like an elf out of Middle Earth, cut his own. He eventually let me in on his secret.

Black locust.

Wiz had figured out through observation and analysis that black locust will behave like a forever firewood tree if you treat it right. By cutting trees that are only three years old, you can harvest wood that is thin enough that it doesn't need to be seasoned or split to burn. Meanwhile, through a technique known as coppicing, the tree grows back from where you cut it. By planting enough black locust trees on site, a seemingly endless supply of firewood can be harvested in a small space by managing the cuts.

I now have a forever firewood hedge planted at my house thanks to the Soul Valley Wizard.

How to Grow

Black locust (*Robinia pseudoacacia*) is a famous tree in Permaculture and agroforestry circles for good reason—it is incredibly easy to grow and serves multiple functions in the orchard as both a nitrogen-fixer, pollinator, forage crop, and versatile timber species. It is a medium-sized tree native to North America but adaptable in temperate climates all over the world (18% of the forests of Hungary have been planted to black locust since it was first introduced there in the 1700s!).[1] Plant anywhere with full sun and expect easy establishment and fast growth—as much as four feet per year. It is even quite resistant to deer pressure because of the sharp thorns that grow along the trunk and branches of the young wood.

Integrating into Your Plant Guilds

There are almost limitless possibilities for leveraging black locust's multifunctionality in a farm system. In a linear guild orchard planting, it can be planted in a repeating pattern to fix nitrogen for the system and provide early shade and biomass before the slower-growing fruit trees grow up. For example, you might plant three fruit trees then a black locust then three fruit trees then a black locust, and repeat.

In larger-scale agroforestry or silvopasture plantings, black locust is often the predominant tree species because it establishes well and is easy and inexpensive to propagate. Unlike most hardwood trees, it can even survive in our bioregion without a tree tube to protect it from deer, although we recommend testing this in your own area before planting a lot of unprotected trees.

Cultivar Recommendations

There are some named improved cultivars of black locust available, many of which were bred for straighter trunk formation for timber and fence post production. The Shipmast cultivar is a well-known selection from Poland. In general, we recommend collecting black locust cuttings or seeds from plantings in your own region and building up your genetic stock that way. One of the main benefits of black locust is how inexpensive and abundant it is, so there is really no reason to spend a lot of money on planting stock.

Propagation

Black locust can easily be grown from seed, which is preferred. If you already have a stand of locusts, especially if it is a genetic strain you want to keep going, you can propagate by rooting live stakes just like willow. If you are growing for fence posts, you will want to prune the trees to keep them growing straight and upright, and these prunings can become the live stakes for your next planting.

Harvest and Uses

Black locust is a great fodder tree for sheep, who will graze the lower branches as the tree grows up and eat the leaves, which have a nutritional value similar to alfalfa. Because it forms a thicket of multiple trunks, it grows back quickly after being cut or coppiced, thus the tree can be directly pruned and fed to livestock or harvested for timber without killing the tree.

Speaking of timber, black locust is the most rot-resistant wood native to North America and is also the hottest-burning firewood. When growing for firewood, consider planting a hedge or windbreak of locust in a line at close spacing (three to four feet). When the trees in this firewood hedge get to the diameter you want for your woodstove, coppice or cut them at the base. They will grow back and in a few years be ready to cut again. If you harvest the hedge in sections on a rotation every year, you will always have enough wood at the right size for a continuous supply. Similarly, timber/post plantings should be close together to encourage straight and upright growth. Also, the robust clusters of white pea-like blooms are edible and a prolific source of nectar for bees. Honeybees produce large yields of high grade, delicious vanilla-almond flavored honey when black locusts are in abundance. Thus they are one of the best (Mid Atlantic) native bee trees along with basswood linden, sourwood, and tulip poplar.

4. CLOVER

Ryan's Story

Communities have royalty. Some royalty exists among the commoners, blending in despite the heavy lifting they are doing. Few recognize their contribution or importance. Clover is like this when it comes to nitrogen-fixers and so was Chuck Kraft and his wife, Mamma Shirley. My parents used to work in Charlottesville in the early 80s. The Charlottesville of the 80s was much closer to a hole-in-the-wall bar than the smug Tesla driver in a Cotopaxi jacket ambience it has now. Before work each morning, they would drop me and my sisters off at Mamma Shirley's house.

Mamma Shirley's house was a rural wonderland in an urban setting. Her yard was a place for bare feet. In the summer, I would walk the rows of tomatoes in her garden with her. She would stop to pick a ripe one, reach into her pocket, pull out a salt shaker, splash the tomato with salt, and hand it to me to take a bite. On any given morning, we'd show up to the house to find a dead bear in the front yard from one of her husband's hunting trips. Or on other mornings, she'd serve us rattlesnake or frog legs for breakfast—with, of course, a can of potted meat and a pile of saltines as an alternative.

Her husband was Chuck Kraft. Chuck spent most of his time as a bricklayer or a fly fishing guide. The man was famous in the fly fishing world; most people who spend time with a fly rod know Chuck as the inventor of the CK special series of flies. Back then, being famous in the fly fishing world was a lot like being a famous farmer—everyone wanted to hang out with you, nobody wanted to pay you.

This was the world that made up my childhood. I experienced a lot of firsts at that house, including my first beesting.

As a small child, I used to love to challenge people to races. One day, just before my parents arrived to pick me up, Chuck pulled up. It was a good day for him on the river. This was obvious because he was full of energy. Before peppering him with questions about all of the fish he had caught that day, I challenged him to a race.

Mamma Shirley's yard was the kind of place where you never wore shoes, or at least I didn't. And nobody ever made me. I was, of course, barefoot on this day.

After discussing the start and finish line, we lined up while Mamma Shirley presided over the competition.

"Ready, Set…Go!" she shouted.

Only a couple steps in, and I stepped on a bee. Chuck didn't stop, didn't even look my way despite my screaming in pain. In what would be an early lesson for me in "dealing with the consequences," he finished the race with a gigantic smile.

After declaring, "I won young man," he walked over to me.

"You OK?" he asked.

Mamma Shirley was holding me as I cried and stared at the bottom of my foot.

She pointed down to the patch of clover where a dead bee lay. She smiled at Chuck.

Chuck said: "Yea, that'll hurt. Good for you though, now you can tell everybody I beat you in a foot race."

Mamma Shirley went on to comfort me while they both smiled at the moment. Later Mamma Shirley would tell me about how good the clover was for the dirt. She said, "It's good for the dirt like that bee sting is good for you." I still don't know what she meant, but even now as I write this, I carry it around as a kind of wisdom.

How to Grow

Clover (*Trifolium* spp.) is a large group of low-growing herbaceous plants that fix nitrogen due to their symbiotic relationship with *Rhizobium* bacteria. While they are native to Eurasia, they have naturalized throughout North America and are a common component of pastures and hayfields in our area of the Mid-Atlantic.

Integrating into Your Plant Guilds

Perennial clover is a great low-growing ground cover for the paths between garden beds or even the aisles between linear guilds in an orchard. In our linear guild rows, we get our trees established with landscaping fabric, but after they are three years old, we remove the fabric and plant a living cover crop of red and white clover. Red clover in particular is a great cover crop for perennial plantings as it is thought to encourage the growth of beneficial fungi in the soil. Annual clovers like crimson clover are an essential part of any short-lived cover crop mix. We often plant crimson clover and oats on vegetable garden beds in the fall for a nice cover that will winter-kill and can be planted directly into the following spring. For walkways and paths, we recommend a low-growing Dutch white clover, which will stand up to foot traffic and drought.

Propagation

Sow perennial clovers any time from late fall through spring. For large seedings, especially in existing fields, you can frost-seed in the winter and allow the freeze/thaw cycle to push the seeds deeper into the soil. Like all legumes, it is best to inoculate clover seeds with *Rhizobium* bacteria to make sure you maximize its nitrogen-fixation characteristics. You can buy the powdered inoculant, mix it into a slightly wet slurry, and toss the seeds in it. Or, if you are lazy like us, or planting a large amount of seed, you can buy seed pre-inoculated.

Harvests and Uses

Clovers' main functions are as nitrogen-fixers, soil stabilizers, and beneficial attractors. Honey bees love white clover so much that we have to be careful walking barefoot on our paths when they are in bloom! Luckily we always have some plantain growing nearby, which when crushed up and applied to a beesting soothes the pain better than any pain medicine.

5. GOUMI

Ryan's Story

I once punched a chicken.

I know how this sounds, y'all. I'm a notch away from entering into the same rarified and ridiculous club as Michael Vick or Kristi Noem. I promise though, that's as far as we will go, no dogs in this story.

My boxing match all started with a goumi bush. Excited about nitrogen-fixing and the taste of those sweet berries after spending some time at Radical Roots farm with Dave and Lee Oneill, I returned to my property and planted a bunch of them. Turns out they are great habitat for free-ranging chickens, which didn't seem like a problem until that damned rooster got a little too aggressive.

When my daughter was a little girl, we thought a charming and formative task for her was feeding the chickens every morning. At first this was a chore she adored. She loved to watch the chickens feed and even picked a few up to pet. The occasional broody hen never bothered her, and for the most part, things moved along smoothly.

One day, this chore took a scary turn for her. While feeding the chickens, our rooster, a big beautiful boy, pecked her hand.

Kai Marley at six years old was a tough little one, used to stomping across patches of thistles barefoot. Not much bothered this kid, but this—this was different. She let out a scream and came running into the house. We did the best we could to console her and reassure her that the best way to handle a bully rooster was to show him you're not scared.

We sent her back out the next day.

That day, my wife and I were sitting at the kitchen table sipping our coffee, feeling like great parents for the moment, when the next thing we know, we see our six-year-old little girl sprinting across the yard barefoot followed by an angry rooster. She was terrified and that was it—there was no getting her back into that chore.

I took over the chicken chores until she was a little older. Goumi shrubs are the perfect habitat for free-ranging chickens. They're not big enough for a hawk to perch on, but they tend to have small thorns that keep away predators.

Being a large adult, I figured the rooster wasn't going to mess with me. This proved true for a short time. I'd go feed the chickens while the rooster hid under the goumis.

It didn't take long for the rooster to get bolder. Soon enough, when I brought the food out, he would charge me. I'd simply step to him, and he'd run back into his hiding.

One day, I made the mistake of turning my back on him after feeding the chickens. As I was walking away, I heard a rush of air behind me. I turned quickly to see the rooster's spurs up at eye level rushing towards my face. Out of instinct, that's when it happened.

I punched the chicken.

That weekend, we brined him in buttermilk and pressure-cooked him, next we paired him with my wife's biscuits smothered with goumi jam. A belligerent rooster often has the same mouthfeel as his pasture posture—aggressive. The goumi jam moistens and mellows the worst of him out.

The problem, after all, is the solution.

Walking among the goumis in my Food Forest, this memory is a fond one. My daughter is 19 years old now, and it's still something we talk about.

How to Grow

Goumi (*Elaeagnus multiflora*), also known as cherry silverberry, is the quintessential Permaculture Food Forest shrub. It produces a delicious red berry, is a natural nitrogen-fixer, is easy to grow in any soil, and is completely pest and disease resistant. This plant truly checks all the boxes! What's more, unlike its cousin autumn olive (*Elaeagnus umbellata*), it does not spread and take over. So if you love eating autumn olive berries like we do but are concerned about its invasiveness, goumi is a great alternative.

Integrating into Your Plant Guilds

Goumi is a fast growing shrub that grows to 7 or 8 feet tall and wide. In our linear guild orchard, where the fruit trees are spaced 18 feet apart, we like to plant one goumi in between each tree. If you are short on space and want to squeeze more in, you can always prune the goumi shrubs back when they start encroaching on other plants. They bounce back quickly from pruning, making them a good candidate for chop and drop fodder for livestock, especially chickens, who love the berries. We have cut a full-grown goumi to the base and watched it grow back to full size within two seasons. Overall, goumi's adaptability and fast growth make it a versatile shrub for many different scenarios including orchard guilds, hedges, and silvopasture. This amazing plant is truly proof of the old saying that, as designers, we are limited only by our imagination.

Cultivar Recommendations

There are only a few goumi cultivars that are widely available commercially in North America, with our personal favorites being the Sweet Scarlet, Red Gem, and Tillamook. There is a lot more breeding work to be done on this plant, especially in terms of larger fruit and smaller seed size.

Propagation

Goumis, like most shrubs, can be propagated vegetatively by softwood or hardwood cuttings, although they are not the easiest plants to root. Although we have not tried it, we have seen them grafted as well. If you are comfortable with grafting and have a lot of wild autumn olives on your property, you could try top-grafting goumi cultivars onto them as a way to get fruit faster and neutralize the invasiveness of the autumn olive.

Harvest and Uses

Of all the berries we grow, goumi is literally our kids' all-time favorite. I think it's because they taste just like candy! We spend hours every summer hanging out by our goumis and eating the berries right off the branch. They do have small seeds that can be spit out or swallowed. If you have the patience, you can collect a big harvest and freeze them and use them for smoothies or make jams. The best fruit leather we ever had was made with goumis. The key to using the berries for things like jam and fruit leather is to remove the seeds with a food mill, similar to the way you would deseed tomatoes to make sauce.

Since there are usually too many goumis for us to eat, we like to cut off whole branches loaded with goumis to feed to our chickens. When we feed them a lot, it even turns the egg yolks red! By the way, the red in the goumis comes from lycopene, a powerful antioxidant found in other red fruits like tomatoes.

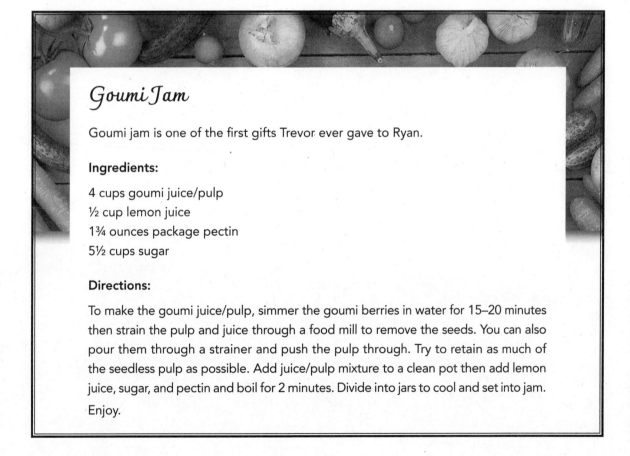

Goumi Jam

Goumi jam is one of the first gifts Trevor ever gave to Ryan.

Ingredients:

4 cups goumi juice/pulp
½ cup lemon juice
1¾ ounces package pectin
5½ cups sugar

Directions:

To make the goumi juice/pulp, simmer the goumi berries in water for 15–20 minutes then strain the pulp and juice through a food mill to remove the seeds. You can also pour them through a strainer and push the pulp through. Try to retain as much of the seedless pulp as possible. Add juice/pulp mixture to a clean pot then add lemon juice, sugar, and pectin and boil for 2 minutes. Divide into jars to cool and set into jam. Enjoy.

CHAPTER 5

Barrier Plants

1. RHUBARB

Nothing gets the taste of shame and humiliation out of your mouth like Rhubarb Pie.

—Garrison Keillor, *A Prairie Home Companion*

Ryan's Story

THERE WAS A TIME in the early aughts when I listened to *A Prairie Home Companion*. It was a soothing, comforting, monotone experience of sonic interest. Keillor had a voice that gave me permission on Sunday evenings to fall asleep in my sitting room chair in front of the fire. Somehow I always managed to wake up whenever he mentioned rhubarb pie. Maybe it was the inflection he used or the way the word rolled off the tongue.

The early aughts was also a time in my life when I had just purchased a piece of property. I've already mentioned a few stories ago that it was the worst piece of property in the county. I'll never tire of saying that part out loud.

In a flash of inspiration brought on by those sleepy Sunday afternoons with Garrison Keillor, I purchased two rhubarb plants and stuck them thoughtlessly in the ground. After all, I grew up Lutheran and probably above average like all those Lake Wobegon kids. I had no idea how to grow rhubarb and even less of an idea what to do with the plant.

A year or two later, I came across Eric Tonesmeier's book on perennial vegetables and along with a visit to Radical Roots Farm in Harrisonburg, Virginia—the home of Dave and Lee O'Neill, my early farming heroes—I began to look at rhubarb as a perfect guild companion in my linear fruit tree guilds. Soon two plants became fifty. Fast forward half a decade, and my farm was rolling in rhubarb!

There are moody plants that take a long time to warm up to you and plants who will make you work for even the slightest signs of affection; rhubarb is not one of them. It is an easy, promiscuous plant. For a few years, I loved the opportunity the plant gave my market farm for a quick, easy harvest to put on the table. During strawberry season, the plant flew off the table!

As we shut down our market garden and I transitioned into teaching Permaculture in the public schools, our farm—now our homestead—was left with an enormous amount of rhubarb that played an important role in the ecosystem. However, at this point, it was still a plant that we hadn't done much experimenting with in the kitchen.

It wasn't until the first party we hosted at our farm post-COVID when we really started understanding the promise this vegetable held. I'm going to safely assume that we all remember the COVID lockdown. Regardless of where you fall on the spectrum of it being a lifesaver or unnecessary, the lockdown was hard. It taught us how valuable our community is to us as humans and how disconnected we've become. As my area of the country was starting to come out of lockdown, in the spring of 2021, we held a rhubarb-themed party. This party featured rhubarb everything. I smoked pork shoulder and marinated it and served it in rhubarb barbecue sauce. We featured a rhubarb margarita as the drink, cold rhubarb soup was on the menu, along with countless rhubarb dessert options including, yes, rhubarb pie. But perhaps my biggest memory was watching my teenage daughter standing over the honey jar and snapping rhubarb stalks into smaller pieces, dipping the stalk raw into the honey for a quick and tasty treat. The culmination of my relationship to this vegetable is best preserved in this memory. Rhubarb offers something for everybody in zesty, effortless, and accessible style.

How to Grow

Rhubarb (*Rheum × hybridum*) is one of the first herbaceous perennials to emerge in the Food Forest in the early spring here in the Shenandoah Valley of Virginia, adding a delightful pop of red and green to the landscape. Other highlights of this plant include its ability to block out weeds, its ease of propagation, and the fact that it is virtually pest and disease resistant.

Integrating into Your Plant Guilds

One of our favorite things about rhubarb is its multifunctionality in the Food Forest. It can be considered a barrier plant, a dynamic accumulator, and a beneficial attractor. We recommend planting several rhubarb around the base of your anchor plant fruit tree along with other barrier plants, approximately three feet out from the base of the tree, to block and shade out weeds. Rhubarb is one of the best plants for this role because it tends to form a natural barrier to encroaching grass and grows large leaves that shade out weeds. In addition, it can be continually propagated and spread to keep weeds in check.

At the same time, rhubarb can be chop and dropped to provide in-place mulch for your Food Forest guild. To chop and drop, let the rhubarb leaves grow through most of the summer and then cut them back at the base of the plant and lay the cut leaves around the soil as mulch. Because rhubarb is a heavy feeder, there is good reason to think that this action will bring nutrients from lower in the soil up to the microbially active soil surface, feeding the rest of the guild. The same technique can be applied to other dynamic accumulators such as comfrey.

Variety Recommendations

There are many varieties of rhubarb, but we like the red-stalked varieties for aesthetics. A great choice is the English variety Victoria. However, because it is so easy to propagate, we recommend finding a friend or neighbor who has a healthy patch of rhubarb and bartering with them for some rhubarb cuttings to get your patch started.

Rhubarb can be planted from seed, but you will most likely plant it as crowns—divisions of the root—pressed into the soil a few inches deep in the fall or spring. Rhubarb is widely adapted in temperate climates around the world (Zone 3 to Zone 8), but does start to struggle in extremely hot areas, so if you are in the subtropics you could experiment with growing in the partial shade of a Food Forest tree guild. It also thrives best in rich, well-drained soils. Heavy wet soils can cause root rot. If you have heavy clay soil on your site, consider planting on raised mounded beds. The mounded beds of a linear guild Food Forest make a great naturally drained site for the plant.

Propagation

To propagate, you will want to dig up the whole root of a plant once dormant (late fall or early spring) and cut it into individual pieces that each have at least two or three dormant buds on them. Side note: this root division propagation technique can be applied to any herbaceous perennial that has multiple growth buds on its underground root or rhizome—think asparagus and comfrey. It is also the same principle at play when you cut potatoes into multiple pieces for spring planting—you are essentially turning one plant into many!

With rhubarb, if you don't feel like bending over to dig up the plant, you can simply take a spade and slice down vertically into the center of the crown, thus cutting the plant in half. Loosen and remove one half and divide that out to create your new planting stock.

Harvest and Uses

Here in Zone 6, we harvest our rhubarb around May. The young stalks of the plant are edible, while the leaves, which are high in oxalic acid, should be discarded as mulch. The stalks have a pleasant lemony flavor and a nice crunch when eaten raw, and of course they can be chopped and made into any number of tangy treats such as pies, jellies, and chutneys. A great tip is to chop and freeze your extra rhubarb for use throughout the year.

If you're a fan of rhubarb and interested in herbalism and medicinal plants, we recommend delving into the fascinating history of the medicinal use of rhubarb throughout the world. It seems to have a common use throughout many cultures as a digestive tonic, and is an important herb in Traditional Chinese Medicine. One fun nutritional fact about rhubarb is that one cup contains 45% of your daily value of Vitamin K![1]

The culinary uses for rhubarb go well beyond the traditional rhubarb pie if you're willing to get creative. We like to make a rhubarb simple syrup for use in summer cocktails, and Ryan has been known to make a delicious rhubarb barbecue sauce. In fact, a rhubarb syrup/reduction can add a nice slightly sweet tang to many savory dishes including soups, salads, and sandwiches, or, Emilie's favorite use, a rhubarb margarita.

2. JERUSALEM ARTICHOKE

Ryan's Story

Farming is hard.

Like chewing a handful of gravel hard.

I have heard Emilie's husband shout many a time the words "Fuck Farming" in the middle of a challenging task that will only be half successful. Logan (Emilie's husband) and I often joke about writing a memoir with that exact title.

But there are perks.

It's all about the food.

During the run we had at Dancing Star Farm, our lunches were the stuff of legend. About an hour before lunch, someone would peel away from the staff working away and head to the kitchen to prepare lunch. Usually, it consisted of what was on hand in the walk-in, fried up in olive oil and sea salt, topped with a fresh fried egg or two. Every now and then Trevor would show up with a pocket full of chanterelles that he'd throw into the scramble, or on great days, we'd catch a rabbit trying to chew on the low tomatoes and turn the little guy into lunch.

One of the great joys I experienced during these epic farm meals was who I got to share them with.

Every meal inevitably featured the storytelling and good-humored conversation of a favorite farmer friend—let's call him Jim.

Jim was the type of human who did the weird thing, the hard thing no one else wanted to, and then reported back so that everybody else didn't have to.

He spent one summer living in a dark, moldy shack just so he could afford to keep working the farm. Jim once showed up to work having chugged a quart of valerian root tea, thinking instead that he was imbibing an energizing and adaptogenic ashwagandha tea for his a.m. brew. He mixed up the plant names. It is recommended to the reader not to do this. Like I said, Jim did it so we don't have to. For those who don't know, valerian root is a sleep aid, and dude was nodding off in the field the rest of the day.

One particularly memorable lunch was also our most spartan. We had a handful of duck eggs and only a mess of Jerusalem artichokes to eat. I was excited about the need to eat these. Farm food is abundant. You don't just harvest one of anything. So we ate without restraint. It was not until later that I remembered the nickname "fartichokes."

We sliced them thin and roasted them with olive oil, dill, and garlic with a little lemon juice added just before the salt.

We feasted that day on sunchokes and duck eggs, and surprisingly this meal left Jim and I feeling full. It was late February, and having finished the farm lunch, we headed out to the greenhouse to get into our late winter/early spring greenhouse flow.

The afternoon was going great. Music playing and both of us working silently in the greenhouse. This is meditative, sacred work, and we were both locked into it. It's the kind of work that hits that pleasure point in your brain when you do it for long enough. Did I mention the greenhouse smelled great? Moist, green, with early basil and thyme and oregano casting their olfactory net. A light soothing rain coming down on the plastic above us enhanced the meditative quality of the work.

That's when the work was interrupted by a large sound that shook me out of my flow.

I looked around disoriented and a little annoyed.

That's when I smelled it.

Jim farted.

A gross, highly offensive smell reminiscent of fermented water chestnuts mingled with duck eggs floated across the greenhouse. I did my best to ignore this obvious breach of etiquette. This was after all a greenhouse. Who farts in a greenhouse? It doesn't go anywhere, the worst kind of hotbox.

Jim's gastric troubles did not, however, end with that release.

For the next hour, they kept coming, at first in drops then in waves. Sounds and smells and through it all we had to stay and work.

That day we discovered where the nickname fartichoke comes from. For a small percentage of people, Jim being one of them, the sunchoke is near impossible to move smoothly through the digestive tract. Instead, the fibrous root travels in a violent pattern.

Be sure to not dine on the sunchoke prior to greenhouse work. Save it for a hike or something.

How to Grow

A perennial sunflower with an edible tuber that is native to North America, there is a whole lot to love about the sunchoke (*Helianthus tuberosus*), also known as the Jerusalem artichoke. The only drawback to this amazing plant is just how expansively it grows! Plant in a sunny location and watch the sunflower-like stalks grow up to ten feet tall, producing beautiful yellow flowers in the summer.

Integrating into Your Plant Guilds

Sunchokes spread quickly through their underground tubers and can take over a garden bed in no time. For this reason, we recommend planting them in a dedicated plot (maybe next to the nettles) and keeping a moat of grass around them that you can mow to keep them from taking over. In a linear guild Food Forest, this could be a strip in the middle of the alleyway, and in an island guild system this could be its own island patch (we don't recommend planting it around the base of the fruit tree as it will quickly shade out most trees). It could also be integrated into a hedge or privacy screen planting.

Variety Recommendations

There are over a dozen varieties of Jerusalem artichoke, including several different colored tubers ranging from white to yellow to red. Our favorites are the Fuseau varieties. One thing to look for when searching for varieties would be ones with tubers that are uniform and easy to scrub clean, as some of the knobbier types can be difficult to scrub clean for cooking.

Propagation

Propagate by digging up the tubers during the dormant season and cutting them into pieces for new plants. This plant is right up there with comfrey in terms of easiness to propagate. In fact, it sprouts so readily that it can be almost impossible to dig out and remove a patch unless you get every single fragment of tuber hidden in the soil.

Harvest and Uses

Sunchoke is one of the most underappreciated homesteading crops, in our opinion. One of the many things we love about it is that the edible tubers can be dug up year-round, even in the middle of the winter if the soil isn't frozen—no need to preserve or store in a root cellar. It is also one of the most dense crops you can grow in terms of calories of food per acre. The tubers can be used in any way that you would cook potatoes and can even be eaten raw. Our favorite way to enjoy them is to roast them in the oven with olive oil and salt until crispy on the outside and soft in the middle (see recipe below). They are also surprisingly tasty when lacto-fermented. In addition to being calorie dense, sunchokes are also high in fiber and inulin, a prebiotic that is a food source for the healthy microbes in the gut (a little too healthy in our friend Jim's case!).

Besides being a tasty food source, the fast-growing nature of this plant presents limitless possibilities for growing chop and drop mulch for your plant guilds and fodder for your animals. We have planted sunchokes in our chicken yard along with several other fodder crops including mulberry, Che, and hazel.

Dancing Star Farm Sunchoke Chips

One of the great things about convincing an actual publishing company to publish one of your book ideas is that it gets you thinking about other book ideas. I've always dreamed of writing a cookbook with just the title *Olive Oil, Sea Salt*.

As a market farmer, we would be asked thousands of times a week how to cook the-specific-vegetable-but-could-be-any-of-them-in-my-hand. Without fail, our answer was always, "Chop up, cook up in a pan, with olive oil and sea salt."

Now guess what this recipe is. You got it!

Ingredients:

Jerusalem artichokes
Salt
Olive oil
Dill, if you're feeling it

Directions:

Chop up the Jerusalem artichokes into thin medallions, put in a pan with olive oil and sea salt, roast at 350 degrees in the oven until done.

On those wild days where we really let our hair down, we add dill. It must be stressed a second time, avoid eating them in a greenhouse.

3. GARLIC

Ryan's Story

I imagine garlic teaches the rest of the alliums what to do. Garlic is a slow burn, and if it was a person, it would strut rather than walk. Garlic is a cross between a battered prizefighter and a peacock. Garlic doesn't make me cry when I'm cutting it up like the onion. When I die, preserve me in red wine and garlic, maybe that was Jim Harrison who said that. For me, garlic is a transition, it's that movement from youth into adulthood.

My wife, Joy, and I lived in Hawaii for five years before returning to the countryside we were raised in. Hawaii, in those years, smelled like garlic and an IPA. We had a restaurant just up the street from us called Ninnikuya; the whole neighborhood smelled of garlic. Bold, memorable, complex, pungent. There is a depth to its experience.

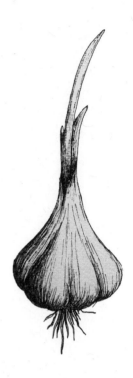

Our last few months in Hawaii were bitter, and they were sweet. We savored every sunset, every morning swim like it was our last in those final days. Our sweet routine of surfing every weekend, every evening, and of course the dawn patrols—where we'd wake up before daylight and be on the water before sunrise—would soon end.

The reality of the move back home finally set in on the last morning surf we would ever have in Hawaii together. Joy was six months pregnant at the time.

She's an athlete—a badass one at that—whose physical ability defies conventional wisdom. Her plan was to surf as long as she was able to. With the baby bump, paddling had become difficult, but she could still surf, so we went.

The day wasn't particularly rough. We paddled out to Diamond Head, a break in town, on a three-foot day. The waves were fun, and everything was going smoothly until I looked up and saw my pregnant wife take a spill.

The fall didn't look hard, and she wasn't anywhere near a reef so I turned and paddled back out to the lineup. When I sat on my board, I figured she'd be right behind me.

She wasn't.

I waited and watched the waves for a few more minutes, but something started bothering me.

I paddled inside to check things out.

Far on the inside, there was blood in the water, and I got scared. I couldn't find Joy anywhere. A moment of panic set in, and I immediately caught the first wave in, paddling furiously because the wave wasn't moving fast enough.

When I got to the beach, Joy was sitting on a broken board, head buried in her hands. I could see blood running down the contours of her pregnant belly. Without looking up she said:

"I think we need to go to the hospital."

On the ride to the hospital, I got the full story. On her paddle out, she was scooting down her board in order to make her belly more comfortable when a wave caught the front of the board, lifting the board up and smacking her in the face.

Her nose was definitely broken, black eyes were setting in, and a gash on the bridge of her nose was bleeding profusely.

It took a minute to wade through the weird optics of walking into the emergency room with a six-month pregnant woman in a bathing suit with two black eyes and a broken nose. The doctors did their finger wagging, and soon enough we were back at our cottage in Kaimuki.

Driving home from the hospital on Waialae Avenue, there was the garlic restaurant, and the whole town seemed to be enveloped in the scent.

At the cottage over a dinner of garlicky shrimp, we both cried a little. It wasn't the pain of the spill she took. Joy was a freak of nature and had been stitched up without pain medication. It was the pain of realizing this was the last time we would surf together in Hawaii. Six weeks later, we flew home. I took the memory of garlic with me along with many other moments. Garlic was in all of those moments.

We returned home searching for more of what we had left searching for. Depth. That fall, having moved home, had a child, and bought land, all in a six-month time span, garlic was the first crop we ever planted. It's the easiest crop, and every year when the family sits around to give the roots hair cuts and braid the garlic that will hang in the kitchen for the rest of the year, we are reminded of the Hawaii years and, more importantly, the transition from youth to adulthood, from adventure to roots. It's a maturation.

Garlic teaches other alliums what to do.

How to Grow

Garlic (*Allium sativum*) is a popular gateway crop for beginning gardeners because of how easy and rewarding it is to grow. If we only had time or space to grow one annual crop, we would choose garlic because we use so much of it in our households and it is such a highly valued medicine.

One of the many things we love about garlic is that it is planted in the fall (October to mid-November in the Mid-Atlantic), which is a nice cool time of year when we typically have more time than the busy spring and summer. It is also easy, and preferable, to propagate each year (see Propagation below), which means that once you get your garlic established, you will rarely have to buy more seed stock.

Unlike some of the other very adaptable Permaculture plants, garlic does need specific growing conditions to thrive, namely rich, well-drained soil, weed-free conditions, and adequate watering. Our best garlic has been in raised beds that have been aerated with a broadfork and loaded with good compost. After planting the bulbs at 6" x 6" spacing, we mulch with organic straw or hay. The garlic bulbs will usually sprout in the fall and continue to grow through the winter, taking off in early spring. Around April, we like to fertilize our garlic with a foliar spray of liquid fish and kelp, which ensures they grow into healthy mature plants later in the summer.

Integrating into Your Plant Guilds

Even though garlic prefers rich, well-drained soil, it can still be integrated into your Food Forest or orchard. One option if you are trying to produce a large amount is to grow your garlic in the alleys between your tree rows. Alley cropping your fruit tree rows with garlic is thought to be beneficial to the orchard system because the sulfur released by the garlic into the soil has antifungal properties that might help combat diseases like apple scab.

Similarly, garlic is a great barrier plant to grow around the perimeter of a fruit tree guild, especially in the early years of establishing the anchor tree. Growing and harvesting garlic involves keeping the soil weed-free and turning over the soil, all of which help young trees fight off competition to get established. The same technique can be employed with sweet potatoes and potatoes grown around fruit trees.

Variety Recommendations

Exploring new varieties of garlic is a really fun adventure as they range in colors, flavor, and texture. After trialing many varieties, we have come to love the sharp flavor and large bulb size of the hardneck garlics, particularly Music, Romanian Red, and German White. Softnecks tend to have smaller cloves and milder flavor but will store longer and can be braided. Elephant garlic is another great option, especially for storage garlic. Although not a true garlic, garlic chives are a great perennial option to get the garlic flavor without replanting every year.

Propagation

Garlic is normally propagated and planted by separating out the individual cloves that make up a bulb. Each clove is planted in the fall to make a new plant. When you are getting started we recommend buying good quality, disease-free garlic from a reputable source and planting extra so you can save seed for the following year. After harvest, we set aside enough seed garlic for the next year and use the rest for personal use. Garlic can also be planted from the bulbils that form at the top of the plant, although this method requires several seasons to develop large bulbs.

Harvest and Uses

Our families each go through about a hundred heads of garlic per year, half of which we use in cooking. The other half we turn into medicine for the winter. Some of our garlic will go into the annual batch of fire cider while the rest is either lacto-fermented in a saltwater brine or fermented in honey. These garlic concoctions are part of our multilayered defense against colds and flus, along with elderberry syrup and astragalus.

Lacto-Fermented Garlic

First, make a brine by combining water and sea salt at a ratio of 3 tablespoons of sea salt to 4 cups of water. Peel your fresh garlic cloves and fill a Mason jar (size of your choosing) about ⅔ full with the cloves. Fill the Mason jar with the brine solution so that all the garlic cloves are submerged. The key to getting the garlic cloves to ferment properly is to make sure they are completely submerged. Take a clean cabbage or collard leaf and fold it up and place it on top of the cloves so that any floating ones are pushed under the brine, then loosely cap the Mason jar to hold it all down. Let the jar sit on the counter at room temperature for 1–2 weeks until fermentation is done. You will see the liquid start to bubble, which will indicate fermentation has begun, and when the bubbling subsides, fermentation will be complete. Put in the refrigerator to store for 6 months or longer.

4. SPRING BULBS

Ryan's Story

Plants come into our lives at unexpected times.

There was a time when I used to love to go to an event in our bioregion known as the Daylily Wine Festival. A famous flower breeder in our area would put on a large celebration every summer where wines from all over the state could be tasted and all manner of conventional landscape plants or flowers could be viewed and purchased. Years ago I spent a couple of years pouring at tasting rooms for several wineries and developed an appreciation for the grape in wine form. From white to red, dry to sweet, I was here for all of the grapes. The festival was great. I wanted to ask questions about how the wine was made and how the grape was farmed. My wife, Joy, just wanted to drink the wine and look at the flowers. I should have known then.

We've seen couples signing up for our Shenandoah Permaculture Institute courses in the last several years. With a few exceptions, we see this general trend as a good thing. Partnerships work well when the partners speak the same language. For those who are in relationships but don't take the course as a couple, we notice a pattern that happens around the third weekend of our PDC.

The pattern is that the individual taking the course starts to feel distant from the people in their lives who are not participating in the course. There are several reasons for this. The community feel of our courses, the time away from each other, the inspired commitment to do inconvenient or hard things in a home system that don't make sense to the partner.

I experienced something similar in my own course. Right around the third weekend, I was feeling distant from my wife. I remember us fighting the morning of that third weekend. I took this energy into the course and had a crisis moment where I was questioning our relationship.

Things didn't improve after the course. I moved with intensity in the direction of designing and implementing a market farm on the property where we were making a home for our family. Joy and I shared some of the values that came along with a market farm but none of the tools for how to get there. Especially, the fact that we differed so much on the values we didn't share. One of these was aesthetics. I wanted productivity and food access, Joy wanted things to look nice.

I kept going. Ignoring a major part of the Human Sector piece, I never considered my wife's interest as a stakeholder and instead kept promising that things would "get better."

In my effort and drive to make a living out of our land, I was alienating my wife not only from me but also from the land. She didn't feel like she belonged. This is the theme we hear in our courses: "We don't belong together anymore."

Fortunately for our family, Joy is a better person than I. She kept tolerating the efforts I was making to grow our property into a farm, even supporting them by helping out at the farmers' market or coming home on a Friday to bag lettuce with me until dark. As a sweet show of support, she even paid for and took a PDC through SPI. The hope was that she would feel more connected to me in doing so. I was excited about this effort and viewed it less as an effort to connect and more as an opportunity to win her over to my perspective on our property. (Y'all see the problems, right?)

Joy was a star student in the course and led her team in an amazing design project. This didn't help in our own system. Now we shared a language, but our project was still something she didn't feel she belonged to.

We talk often about finding leverage points in design. The guiding question along these lines is, "What small action can I take that will make the biggest impact?" The aha moment for me came just last year.

It was our 15th course at SPI, and the 25th year we had been together. This year, for the first time ever, we were using our property as the design site for the course. We try to train our students to think about any tension that is noticed in stakeholder interviews and wonder how it might show up in the design and how it can be resolved. This line of thought can be gotten at through the client interview.

Joy and I were being interviewed by the class, and things were going great. That's when they asked me what Joy's favorite flowers were. I had no idea. Next, they asked me if she had a favorite color of flower.

My wife, joined by the class, all directed their attention at me: crickets.

Just as things were getting uncomfortable, Joy jumped in to save me: "He doesn't know this, and has never asked: I love yellow flowers, especially the ones that first show up after winter."

Boom.

I was embarrassed.

And I had my leverage point.

This became a running joke for that particular course, that Ryan doesn't even know his wife of 25 years' favorite flower. I went to work that fall planting as many bulbs and flowers as possible.

As I'm writing this, I'm staring out of the window at the kitchen garden. It is April and the garden is alive with yellow. Daffodils, tulips, which will morph into the summer to daylilies flashing yellow. This has become Joy's favorite morning coffee spot and a simple resolution to the conflict brewing in our experience of our landscape. It's the place that she now has ownership over and feels that she belongs. It's a place that I've become interested in as well. For years I dismissed these bulbs and daylilies as conventional fare that populated boring suburban landscapes. This much is true, however, the vibrancy and joy that these plantings have contributed to our unconventional landscape are priceless. Through my wife's experience of belonging to our landscape, I've developed a close relationship to the perennial flowers, and am grateful for the impact they have had on the most important relationship I have.

How to Grow

There are many important functions that need to be addressed in a guild planting, but one that is often overlooked is beauty and aesthetic enjoyment. That is where the spring bulbs come in. These are a group of perennial flowers that can be planted as bulbs in the fall and will bloom throughout the spring and early summer, before many other plants have flowered. They include the classic spring flowers such as daylilies, canna lilies, daffodils, and tulips as well as many others. Centuries of intense breeding have gone into these ornamental flowers, making them easy to care for and widely adapted. In addition to being beautiful, they can also function as barrier plants, beneficial attractors, and an income source through cut flower sales.

Integrating into Your Plant Guilds

To stack function in our guilds, we utilize spring bulbs as barrier plants to choke out weeds and grasses. Like other barrier plants, we aim to line the outer perimeter of our guild planting with closely spaced (two to four inches) bulbs. Daffodils are the classic barrier bulb and are perhaps the easiest to grow and perennialize. Other great options include anything in the lily family, including daylilies and Persian lily, as well as tulips and irises. For deer-resistant bulbs, we recommend hyacinth, crocus, gladiola, crown imperial, allium flower, dahlia, and guinea hen flower. Peonies are a must-have, not only for their wonderful aroma and economic value as a cut flower but also because they are thought to attract Tiphia wasps, a beneficial insect that may help control Japanese beetle pest populations.[2] In our orchard, we have planted peonies throughout every few guilds for this purpose.

Variety Recommendations

It is really hard to give variety recommendations because there are so many and most of the time the choice comes down to the color you are looking for. For daffodils, we really enjoy the pale yellow and two-toned varieties over the classic yellow. Daylilies bring a gorgeous pale orange hue that is not common in other flowers, and certain tulips can offer unique splashes of deep purple and red. If you are in a warm climate (Zone 8 or above), you can grow the edible canna lily (*Canna edulis*) as a perennial.

Propagation

In temperate climates, plant the bulbs in the fall for a bloom the following year. Some plants, like irises and daylilies, can be divided by root cuttings. Daffodils, tulips, and other bulbs that do not divide as easily can be dug up and replanted from a friend or purchased at a nursery.

Harvest and Uses

All of the spring bulbs are valuable cut flowers that can be a supplemental income to the farm and a great gift. In terms of edible bulbs, the most common and delicious would be the daylily (*Hemerocallis fulva*), a native of Asia that is naturalized in North America and can be found growing wild. This is the classic daylily with the orange flowers that open in the morning and close up by night. The shoots, flowers, and bulbs of this plant are all edible. Please note that there are many other flowers with lily in the name and most of these are not edible.

Our favorite way to enjoy the lily flowers is by frying them up into fritters, almost like a lily flower tempura. For a more substantial dish, you can harvest the bulbs during the dormant season and cook them just like you would fingerling potatoes, boiled and sautéed or roasted in the oven with olive oil and salt.

5. ELDERBERRY

Ryan's Story

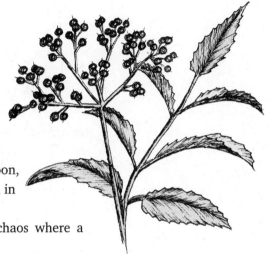

The knife fight we got into was resolved quickly. There is always a winner.

There is no standing on business in the garden. Two hours into cultivating the rows of peppers on a 100-degree day and office niceties like careful language, trigger warnings, and filters melt away. It's part of the charm, really. Intimate stories are shared too soon, and when someone is messing up, you tell them, often in four words or less with colorful four-letter words.

The hierarchy is flattened. It's a kind of social chaos where a rhythm takes a bit to grasp hold of.

When it's good, it's magic, like fingers on a hand moving through the day, when it's bad, it's a dumpster fire.

Part of me loves this—friendships are developed fast in this setting. Cream rises on the farm—either you have it or you don't. The market table is the great evaluator. In the workaday world, behavior is rewarded over product, in the garden, on the other hand …

It's also a setting where fights take place partly due to the pressure from that market table, partly due to that rescinded professional filter some of us move through our days with.

Voices rise, people flip out, maybe it's the heat, or the wind—communication just gets hard. This was the setting for my one and only knife fight with my friend Jenny Taylor-Jones.

One day, over a poorly germinated bed of carrots, Jenny pulled a knife on me. Except that it wasn't a knife, it was the Hori Hori I got her for a wedding present.

And if I am to be a reliable narrator, she didn't exactly pull it on me either, it was already in her hand as we were digging out thistles barehanded.

Jenny and I shared the experience of having both been college athletes, both being trained in Permaculture in the same PDC, and both having been colleagues in the same counselor training course in grad school. Now we were farming together.

We loved each other, but at the moment, shit was getting tense. As I write this, I don't even remember what we were arguing about, probably blaming each other for the poor carrot germination.

Both Jenny and I came from local social cultures that had an edge to them. Our edges came out that day. Voices rising in the field, eyes ablaze with fury, me wanting to be right, Jenny wanting to be right. That's when Jenny shook the only thing she had in her hand at me to go with an expressive tongue lashing. I caught myself and immediately decided a better tactic was to start fighting unfairly. That's when I said, "Please don't threaten me with the knife, Jenny." In my best counselor voice.

Ahh, there it is—the fabled counselor's voice. It's the thing that is fighting words to other counselors. It's a firm, paced way of speaking meant to communicate calm, but in the wrong contexts can be weaponized for condescension. Ask any kid of a mental health counselor, they hate the counselor's voice. In this instance, I didn't feel threatened, I wasn't worried about my safety, I just wanted to win the argument. Jenny caught it immediately.

She chucked the Hori Hori at the ground and growled: "Ryan Blosser, don't use your counselor voice with me."

She was right.

We finished our task that day, but the farming project failed in the weeks to follow due to other tragedies and poor communication. We each moved on to other projects, years passed without us talking about it.

Enter elderberry.

Dust settles and what is left is a sharper view of everything. A few years later, Jenny invited me over to her house to tour her garden. She was proud of the little Permaculture paradise she had developed on her urban lot in the city. One of the early successes she had was with two of my favorite elderberry varieties: Black Lace elderberry and Emerald Lace elderberry.

During the tour, she led me back to a compost pile where a thriving Emerald Lace had established itself from a discarded pruned cutting.

Jenny knew that I needed an Emerald Lace to pair with my Black Lace. She dug it out of the ground with her Hori Hori—the same Hori Hori—and handed it to me as its root ball spilled soil.

"I want you to have this."

As I'm writing this, I'm looking out my kitchen window at the Emerald Lace elderberry she gifted me. It's perfect in its place.

In plant lore, the elderberry is the great teacher, the healer, the mentor. The connector, who teaches the rest of the herbal garden how to be their best selves. I love this about the plant.

I see Jenny now but not enough. Ours is the type of relationship where we will text each other out of the blue just to say we miss one another. As adults, it's hard to weave our lives in and out of each other's while balancing work, farms, writing, parenting.

Like the elderberry, she is a constant and a healer. When the hell-broth of tragedy descended onto my life in the form of a horrible and very public vehicle collision years ago, I had people texting and calling to ask if I was alright. But it was Jenny who showed up.

We didn't talk about anything when she came over, the trauma was too fresh. We worked. She knelt in the soil with me side by side, and we pulled carrots. It's the elderberry and it's Jenny that reminds me to be a better person. I love them for it.

How to Grow

Elderberries (*Sambucus* spp.) grow in temperate climates around the world, with many different species native to both North America and Eurasia. A medium-sized upright shrub (six feet tall by six feet wide when mature), it produces edible and medicinal flowers and berries and serves as a gorgeous insectary plant and native pollinator.

In North America, our native species is *Sambucus canadensis*, which grows naturally along forest edges and thrives in riparian zones. Because of its native habitat of partially shaded, moist microclimates, elderberry does not do well in extremely hot, dry locations. On a small home scale, we recommend planting in the partial shade of a guild, along the edges of the garden, or near creeks or tree lines. It grows fine in open sun; however, if you are growing for commercial production, you will want to make sure you have drip irrigation available in case of drought. The easiest way to prune *Sambucus canadensis* is to cut the entire plant down in the fall. The following season, the plant will produce new vigorous canes that will flower and fruit in mid-summer and be less susceptible to pests like borers than over-wintered canes.

In our bioregion of the Shenandoah Valley, we are fortunate to have a small no-spray elderberry farm that specializes in growing and selling elderberry products. The Long Acre Farm, run by Cody and Paige Wilmer, has been an invaluable resource in helping us learn how to grow elderberries in our specific climate and soils, and we recommend finding your own local growing experts, buying their products, and learning from them as much as possible.

Integrating into Your Plant Guilds

According to Cody, elderberries are relatively shallow rooted and therefore need consistent moisture and a weed-free root zone for best production. Cody has observed that they grow naturally in his area along with blackberries and imagines that on a small scale you could create a living "fedge," or fence hedge, to discourage deer by tightly interplanting elderberries and blackberries along the perimeter of a garden.

Cultivar Recommendations

At Long Acre, Cody grows mostly improved cultivars of the American *Sambucus canadensis*, but he also grows some European *Sambucus nigra* as well as hybrid varieties that are crosses of the two. The biggest challenge to growing the berries tends to be keeping the birds from eating them before harvest. Cody has had success managing birds by planting varieties such as Ranch and Marge, which have berry clusters that droop to the ground and become hidden from the birds. Another pest, spotted wing drosophila, can be managed by planting early-ripening European and hybrid varieties. Ryan's favorite varieties are the Emerald Lace and Black Lace elderberry, which he has planted around his front porch so that he can enjoy their striking beauty and evocative colors. Other outstanding varieties for berry production include Bob Gordon and Wyldewood.

Harvest and Uses

Once you've grown a nice stand of ripe elderberries, the next step is harvesting and storing away the berries. On a small scale, Cody recommends snapping off the entire head of ripe berries and placing them in a paper bag and putting them directly into the freezer. When you are ready to use the berries, shake the berries in the bag to loosen them from the stems and then sift through a small hardware cloth or screen. Most people try to separate the berries completely from all stems and leaves before use due to concerns about toxicity in the plant material.

Next, you want to extract the juice for use. One option is to cook the berries down in water and then strain them out. This will extract the most nutrients. Another easy option is to use a steam juicer to release the unconcentrated juice. Either way, once you have the juice, you can freeze it for later use or immediately turn it into a final product. It is nice to keep some pure juice on hand throughout the year to add to cocktails, beverages, syrups, etc. Long Acre Farm makes an elderberry lemonade that is the most refreshing thirst quencher on a hot summer day. We like to make a large batch of immune-boosting elderberry syrup each summer, including local honey and ginger and turmeric grown on the farm, and then freeze it and give it as gifts during the holidays.

Emilie's Elderberry Syrup

Ingredients:

3 cups fresh or frozen elderberries, de-stemmed
4 cups water
2 tablespoons fresh ginger, minced
2 tablespoons fresh turmeric, minced
1 cinnamon stick
¼ teaspoon black pepper
1 cup local honey
¼ cup lemon juice
Organic vodka or other 80-proof spirits (optional)

Directions:

Combine all the ingredients except for the lemon juice and alcohol in a large stock pot and bring to a simmer. Simmer until the liquid is at least reduced by half, or longer for a more concentrated syrup. Add honey and stir. Strain out solids and pour liquid into Mason jars for short-term storage in the refrigerator or long-term storage in the freezer. If you are storing in the fridge and would like to increase how long it will keep, top off each jar with 80-proof spirits.

CHAPTER 6

Beneficial Attractors

1. TULSI

Ryan's Story

Nick, my good friend and farm manager—was bugging me for most of the day.

While cutting the lettuce, he asked, "Can we plant?"

While washing radish bunches, all I heard was, "I really think it will sell well."

We were early in the season, and for probably the only time all year, we were ahead on our harvest list. It was a Friday just before the market, and we usually wrap things up in the walk-in and head to the deck for a happy hour drink sometime around dark. Today, we were finished, and it was not even 4 o'clock.

So I caved.

Nick noticed an extra bed near the strawberries that somehow didn't get onto our season plan. Meanwhile, we had 50 tulsi plant starts in four-inch pots left over from the early season plant sale. I was slightly familiar with the plant, but this was Nick's thing.

He was enamored of the plant in the way a young farmer gets excited about a plant after they hear an old farmer talk lovingly about it at a conference. Nick hadn't heard about tulsi from an old farmer, however—he heard about it first from Trevor, and Trevor, especially in his 20s, has always been a little vulnerable to magical thinking.

"Did you know it can help you adapt to anxiety?"

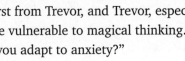

"It goes great with lemon balm as an iced tea on a hot day!"

Because of Nick, I knew this plant without ever actually knowing this plant.

We spent the rest of that day planting out the tulsi with me doubting that it would sell, doubting even the adaptogenic effects everyone kept talking about. At the time, I was a bit hardened to the hippy shit that seems to have enveloped plant talk. I was more focused on the practical market potentials of whatever plant would be taking up valuable real estate on the farm.

Tulsi? Ehh, it smells funny and takes too long to harvest.

But strawberries? Now they'll sell!

That evening, I went to work as was my custom. A start-up market farm is not a financially viable project for a parent, and I had two children so I worked nights at the rehabilitation center supporting youth with disabilities. My shift was from 4 p.m. until midnight, and then back up to farm the next day. Those seasons were long.

On this night, everything seemed fine until I got a text from my wife, Joy, that read,

"I'm not feeling too well, it came on fast! Can you come home early?"

This was a big ask for my wife. She grew up in a household where they made their kindling by chewing the bark off nearby trees. When times were real tough, they pulled out their own fingernails to burn for heat. She is the toughest person I've ever known and to ask for my help was veering way off-brand.

And we were short-staffed.

An hour later, I got another text: "I think I'm alright, but I just threw up."

I started to worry and immediately tucked the worry into the back of my head.

After the students were in the dorm and it was quiet time, I called home to check in. She let me know that she was feeling better now that she was laying down.

We hung up. I kept worrying and watching the clock. By 11 p.m., I was the only one left on shift, and there was no way I was getting out of there before midnight. So I just hoped she was fine. Summer is not a time when my family tends to be overcome with things like the flu; this was out of pattern.

Around 11:30, another text came through: "I'm not ok. But don't worry."

I called immediately.

"I passed out going to the bathroom, and I think I hit my head but I can't remember anything that just happened. My head hurts, and I'm laying on the floor."

I talked her through what was happening, whether I should call an ambulance or not. During the phone call, she got herself back into bed and was feeling better.

The minute my relief showed up, I ran out the door and hustled home. I walked into the house to find her laying face down on the cold bathroom floor with blood, vomit, and shit all around her.

I helped her sit up and got her back up to bed. She couldn't explain what was happening to her. At that moment, her eyes rolled back into her head like something out of a horror movie about demon possession.

I put new clean clothes on her and got her into the truck. We live way out in rural Virginia, so calling an ambulance would be much slower then me just driving her, plus we were broke and an ambulance ride is expensive.

"I'm going to get Tamayo," I told her, and ran back into the house for our three-year-old. I told him we were going on an adventure and scooped him up.

On the way out the car, he asked,

"Daddy, what's that smell?" There were a lot of smells in the house on account of Joy's loss of her own physiological control. "You," he said. "You smell weird." I had no idea what he was talking about and tucked the thought away. I got him into the vehicle, buckled him up and off we went.

Before speeding out of the driveway, I took a minute to breathe. That moment, I remembered Nick and I had spent the day planting tulsi. I reached my hand to the back seat to my son and asked,

"Is this what you smell?"

He nodded yes.

I didn't know if things would be OK or not at that point. A lot still had to happen. But that moment, that smell, with my wife passed out in the seat next to me, and my son smiling at me like there wasn't anything in the world to worry about—that moment it didn't feel like things were going to go south.

I sped away.

There is so much more to this story. I'll save it for another time. Two days later, my wife woke up in the hospital with a smile and a question: "What happened?"

It turned out she was pregnant (we weren't even trying). Because we weren't trying, nobody knew or caught the ectopic pregnancy before it was too late. The egg was stuck in her tube and essentially put a hole the size of a pencil eraser that caused her to bleed half of her body's blood out into her stomach. Her recovery was as fast as her decline.

Two days after getting out of the hospital, she was running four miles a day. For her recovery, I made her tulsi tea every night and started drinking it myself. That summer was the summer I became obsessed with it. It is paired with an exciting and traumatic event on the one hand; on the other hand, because of how this plant paired with that day, the smell relaxes me. Or maybe it actually is adaptogenic. I don't know what to believe.

It never sold well that summer at market, and I made it a point to let Nick know every chance I got.

How to Grow

Tulsi, or holy basil (*Ocimum tenuiflorum*), is the most fragrant and delightful herb most people have never heard of. A member of the mint family and closely related to basil, it is an annual that thrives in warm temperatures and full sun.

Integrating into Your Plant Guilds

We think tulsi should be a part of every medicinal and culinary herb garden. Although it is an annual, it is a prolific self-seeder, which means if you plant it in your guilds and allow some of it to go to seed, you will have tulsi seedlings popping up the following season. Self-seeded annuals are a great way to get an effortless yield in your guilds; in fact, you could make an entire self-replicating guild out of them. Some of our other favorites that self-seed regularly and would be good companions with tulsi include cilantro, dill, poppies, calendula, borage, calendula, and marigolds.

Variety Recommendations

There are at least three strains of tulsi that have been bred, and all are worth planting for their slightly different characteristics and medicinal benefits, including Vana, Krishna, and Rama.

Propagation

Start your tulsi seeds like you would common basil and transplant out after the danger of frost has passed, or direct sow into your garden in late spring or early summer.

Harvests and Uses

Tulsi tea made from freshly picked tulsi leaves (before they go to seed) is a truly wonderful and calming beverage. It is so flavorful and aromatic that it can be brewed by itself, but it can also be combined with other calming herbs such as lemon balm, bee balm, and mint. Just like common basil, you can harvest the top portion of the plant multiple times during the growing season, and it will continue to grow and branch out. By practicing this continuous harvest method, you will get more yield out of your plants than if you harvest the entire thing at one time. Speaking of yield, tulsi is prolific. So be prepared to dry and store a bunch for tea to enjoy throughout the dormant season.

2. ZINNIA

Ryan's Story

Due in large part to a life spent intensely following my interests beyond convention, I have been fortunate to spend a great deal of time with both farmers and mental health therapists. At first glance, there is little shared between the two, but a deeper experience of the professions and a peek behind the curtains leads to an inescapable observation about both. Counselors and farmers are, on balance, gigantic goddamned lunatics. One nurtures the land, and the other nurtures people's psyches. This world of ambiguity and nuance they both splash around in cultivates a sense of grounded chaos. An ordered looseness that can pass for a stoic calm, but don't get it twisted. Most of us are full-on loony.

My life has been saved by many counselors. This story is about one particularly important woman who came into my life simultaneously with the great zinnia and left too early, leaving a ritual of planting zinnias and gladioli in her wake.

Death is a thing—a constant thing. There is no arguing, no negotiating with it, and deaths of despair are no different.

In 2011, while in the middle of a start-up farming collective project at the time named Dancing Star Collective—which would morph into my farm project Dancing Star Farm—our friend and collaborator took her own life. Our community and farm project were rocked by this event. Friends lost a mother, a wife, a colleague. There are moments that take us all to our knees. Some have several. It's coming for you, it's coming for all of us.

I can trace several consequences to this event. The start-up failed shortly after, the friend group disbanded for a time, we all found our own ways to cope. I farmed hard and threw myself into the work of building a nonprofit. But things weren't quite right.

That was when I decided I needed help again, so I did what any mental health clinician would do in this situation—I called a therapist for help. Brené Brown has a line about seeing a therapist who sees therapists. She says, "We have to go to therapists who see therapists because their BS meters are good." We are, after all, a slick lot, trained in the Jedi way and able to duck and dive any support offered through talk therapy.

Harriet was a great woman and professional counselor. I was fortunate enough to develop a therapeutic relationship with Harriet on the client's side of things during my training. Not everyone can be a counselor, and often those who can have a great deal of work to do in order to get there. Harriet worked with me for two years, helping me undo the brash competitive swagger I had developed as a college basketball player that happened to be getting in my way in the therapist's seat.

During the two years I worked with Harriet, she supported my development through grief, problems with my marriage, interpersonal relationship difficulties, and the overwhelming problem of my oversized ego. It was a lot. And I loved her for it.

Most therapy is done on a couch—I know, a cliché—but my favorite session with Harriet happened outside in July. I was having problems in my marriage. Joy and I weren't getting along. Things had gone sour, as they so often do.

Harriet was blunt with me. "It's your fault."

"What?" I asked.

"You have stopped trying. What is it that you want from your wife right now?"

"Attention."

"So then give her attention."

She then took me on a walk in the garden. Zinnias were showing off in full bloom. That and gladioli. She zipped through the garden, cutting flower after flower and explaining to me how she first learned to grow flowers as a young clinician in rural Virginia. Her interest in flowers was how she started to earn the trust of the women in the community she was working in.

She sent me home from that session with a directive: "Give your wife the attention you want from her," and concluded the session with this gem:

"When things are falling apart, try flowers and morning sex."

Needless to say, it worked. Joy and I got through our rough patch, and I started growing zinnias.

It was this memory of healing and help that nudged me to reach back out to her when I needed more support getting through my friend's suicide. We set a date to start working. The work would be clear: grief is not easy work, nor is it simple, and I knew I would be in good hands.

That's when I read the news report. Harriet had been found dead, alone, in her car in the rural town where she got her start as a young clinician. It was a suspected suicide.

I thought of our work together, I thought of the work we wouldn't get to do. I had questions; none of these questions would be answered.

Then I thought of zinnias.

Bright, psychedelic, and a little melancholy in that way a perfect spring day is melancholy. The kind of day that is so beautiful it almost hurts to drink it in. You know the day is fleeting, that a storm will roll in soon or the wind will kick up, but you look in anyway, try to savor it. It's the bloom on the zinnia, long-lasting, but still fleeting. It's a comfort to know it will come back every year. Even though you know it will also leave you.

How to Grow

Zinnia (*Zinnia* spp.) is an annual flower in the aster family known for its beautiful, long-lasting summer blooms. While we've been growing zinnias for years, we wanted to learn more about this plant from a professional flower grower, so we talked to our friend Susanna Byrd of Spring Creek Blooms in Charlottesville, Virginia. Susanna grows flowers using ecological methods that prioritize the health and balance of the soil, insects, and wildlife of her farm community. According to Susanna, zinnias are a great flower for beginners because they are so easy to grow and resilient to heat.

Susanna starts her zinnias from seed in late March in her climate and plants them out after about a month. One thing to know about zinnias, she says, is that they are definitely heat loving. While they can survive a light frost, they should really be transplanted out after the last frost, just like tomatoes and peppers. For Susanna, they will start blooming in June and continue for at least a month if not more. Once the plant is about ten inches tall, she will pinch off the top growth to encourage the flower to branch out and develop more blooms. Although she does trellis hers with flower netting, she stresses that, on a small scale, it's not as important and that some varieties are bushier and more sturdy than others.

Integrating into Your Plant Guilds

Zinnia's primary function in a guild, other than beauty, is as a beneficial attractor. Susanna says they are particularly popular with butterflies and hummingbirds in her gardens, especially the red and the brighter varieties. On the flip side, they seem to attract Japanese beetles, so perhaps they can be used as a trap crop to keep that gnarly pest away from the rest of the orchard. You can transplant them out into your linear or island guilds, and if you're lucky, they will self-seed in the years to come. It is also worth trying to direct seed in your guild if you have bare soil, or mixing into a beneficial insect mix for larger-scale direct sown plantings.

Variety Recommendations

In terms of varieties, Susanna recommends Oklahoma and the classic Benary's Giant. For something unique, she is a fan of the varieties bred by Floret Flowers in Washington State, which include some beautiful muted colors like Dawn Creek Peach and Golden Hour.

3. CANNABIS

Ryan's Story

I was halfway through the row when the helicopter showed up. It was 2010 and a group of friends and I had taken the leap from homestead scale to growing mixed vegetables for a CSA. I was excited to finally be farming. This was the fruition of a dream that started in 2004 when we purchased the property. That summer, I had just purchased my first stirrup hoe from Johnny's and had grown fond of the meditative work of cultivating with it. I ignored the helicopter—until I couldn't. The push, pull, and crumble of the earth, normally a joyful relaxed moment, grew irritating when I realized that the helicopter not only was not leaving but was hovering directly over me. I couldn't take the noise or the wind pushing down from the blades throwing freshly cultivated topsoil into a cloud of dust surrounding me. I went inside to wait and watch for what would happen next.

There was a moment of relief when the helicopter flew off. I stepped outside to get back to work on the okra patch. It was 2 p.m., and the sun was bright. I watched the helicopter fly to the top of the hill where the road crests. Just as my shoulders were dropping and my mind clearing from the confusion of why a helicopter was hovering over me while I worked in my field, I noticed the chopper turn around at the top of the hill and zip down the road low to the ground, this time leading a caravan of unmarked black SUVs followed by one state trooper car with its lights spinning.

I almost shit myself. I started chanting as I watched the convoy speed down the road, "Not my driveway, not my driveway," and that's when they turned into my driveway. The helicopter flew directly above me, and I watched with my mouth open in pure confusion and terror as each vehicle rushed down my driveway. They were in a hurry. I even saw one vehicle catch air on a small bump in the driveway.

The choreography was impressive. They slammed their brakes, and the car doors all opened seemingly simultaneously. Fifteen men in camouflage fatigues spilled out, some kneeling, some standing, all pointing guns at me.

A little fella ran at me, stopped short and shouted: "Mr. Blosser, drop your weapon." I let the stirrup hoe fall to the ground.

"I'm scared," I responded.

"What do you have to be scared of? Do you know why we are here?"

The little guy had to jump just to get face-to-face with me. His aggression worried me, and I remember thinking how relieved I was that he didn't have a gun pointed at me. This guy would have given coffee the nervous jitters.

The question he asked lingered in the back of my head. If you've seen the movie *Goonies*, there is a scene where the character Chunk is told by Mamma Fritelli to "start from the beginning." My entire history of rule breaking flashed through my head.

That time I bought snacks for my friends with my parents' gas card and didn't tell them? That time in kindergarten when I swapped my regular milk with my friend's chocolate milk and pretended that nothing happened? The toilet paper I stole from my work that year we could barely keep up with our mortgage payment? Or the pen from the bank I kept just last week when I made the farm deposit? All of it came flashing through.

I said the only thing I could think of to answer his question. "I'm scared because you all are pointing automatic weapons at me."

That's when a man in a suit and Cool Hand Luke-style sunglasses stepped forward and shared that they were here to eradicate the marijuana I was growing, and that if I stayed calm and cooperated, everything would go smoothly.

This got the little guy excited. He jumped up and down and said, "Try something." I looked in the direction of the suited man. He asked, "Can we take a look around your property?"

I said yes.

That's when the little fella said, "No, you first."

Fifteen men marched me at gunpoint to the back of the property to search for what they claimed was marijuana. Anyone watching would have thought my little farm in Virginia had just turned into Afghanistan.

My first worry at this point was that some neighborhood kids had hid their plants in the back of my property.

Soon, the scary turned funny. As 15 men fanned out literally ripping up tufts of grass in search of this plant, I realized that they had made a mistake. Thirty minutes into the ordeal, the men fell apart. The helicopter still hovering over the scene was shouting down through a bullhorn, "The plants are right there!"

Meanwhile, the men on the ground were starting to fight amongst themselves nervously. It wasn't until I heard one of them shout at the helicopter, "It's two o'clock in the afternoon, take off your fucking infrared! You can't see shit!" that I started to openly laugh.

And the little man who, moments before, stood taut and ready to pounce, had gone slack. He was now standing next to the man in the suit with the body language of a ten-year-old who just lost the ice cream off the top of his cone.

It took another 30 minutes for this hilarious shit show to wrap up. On their way out, I offered the suited man some tomatoes and okra for the road, thinking he would at least apologize for the terror he had just put me through. He declined and said, "We just gave your neighbors something to talk about."

They left a little after 3 p.m. that day in August. For over a decade, the terror of the moment lingered. In fact, to this day, I still can't hear a helicopter overhead without the thought "*They* are coming." While I can fill a book about the lessons learned that day about our law enforcement strategies and personal rights, the experience gifted me love. It tied me to a plant I had never grown but soon became intensely interested in. If our government can afford to spend that much money and time and equipment ripping up a small farmer's property on bad intel over a plant, then this thing must be special. I started reading everything I could get my hands on about the plant, and the second CBD hemp became legal, I got a license to grow it and began a much more intimate relationship with the plant. Soon full legalization in Virginia followed. I have since developed a love for growing the plant and appreciate that now not only is it legal to grow here but that every time I hear a helicopter, I can relax. They aren't coming for me.

How to Grow

We have only been growing cannabis on a small scale since it became legal in Virginia, so we caught up with our friend and cannabis grower Ben Samuelson to learn more from a seasoned expert. Ben and his wife, Betsy, run Seed and Soil Farm in Monroe, Maine, where they specialize in cannabis seed and breed for outdoor, natural growing methods.

Cannabis (*Cannabis* spp.) is a notoriously widely adaptable genus, but if you want to grow a high-quality smokeable or extractable flower, there are a few important considerations to keep in mind. In general, the plant loves warm growing conditions, similar to corn or tomatoes. Ben recommends starting seeds indoors in a protected nursery setting (68 degrees is ideal) in a sterile seedling mix. Start your plants no earlier than two weeks before the last frost and then transplant them into the field before they get rootbound (seedlings grow fast!). Seeds germinated as late as mid-June will have no trouble reaching maturity and producing many ounces or more than a pound of dried flowers, so feel free to take a shortcut and germinate in a half-pint-sized pot when peppers get transplanted into the ground. The cannabis will still end up being the biggest plant in the garden! In colder climates, they may benefit from growing in an unheated high tunnel, but that's not necessary in most of temperate North America with appropriate variety selection.

The biggest growing challenge, especially in humid areas, is getting the flower (buds) to ripen for harvest without developing Botrytis mold. To prevent this, you will want to give your plants as much airflow and open sunlight as possible and plan (or hope!) for a ripening window of dry weather. One way to increase the odds of dry harvest weather is to plant multiple strains that ripen at different times. Cannabis is also susceptible to a bacterial disease called Septoria leaf spot; however, Ben says this can be managed by selecting the proper genetics (see variety selection below). Other than these issues, the plants are very pest and disease resistant when grown outdoors in healthy living soil. Indoor growing, which we don't recommend if you have access to soil and sunlight, presents its own sets of interventions and challenges that are beyond the scope of this book.

Integrating into Your Plant Guilds

Ben likes to place his plants for home use in his "Zone 1" kitchen garden close to the house so that he can enjoy the beauty and aroma of this amazing plant. One year, he said, he planted lettuce around the base of his young transplants, which he harvested by the time the cannabis got large. In a Food Forest guild, cannabis plants fill the niche of any herbaceous low-growing plant, with the extra consideration that you want to put them in full sun and in the best place to get airflow. If you start thinking about growing cannabis for one of its many uses other than flowers, your growing strategy might start to change, and you have options that will only be limited by the legal restrictions in your area. For example, if you have access to inexpensive bulk seed, it could be a great self-seeding forage crop for a low-input chicken production system. One thing Ben wants to stress though, related to cannabis growing etiquette: if you are growing non-feminized seed, be careful not to let males mature and shed pollen lest they interfere with your neighbor's grow.

Variety Recommendations

Favorite Strains

This is a big and very subjective question, but as a seed breeder himself, Ben does have some recommendations. First, he says you can't go wrong with the classic Blueberry Muffin, an early variety that "seems to be everybody's favorite strain" and literally smells like blueberry muffins. Ben is also a fan of Humboldt Seed Company's Jelly Donut, which ripens even earlier than Blueberry Muffin.

For a stellar late-season variety, Ben recommends Omaha Jazz, which he bred on his farm and is available through Seed and Soil of Maine. This strain has high vigor, a unique terpene profile, and ripens around Halloween. Most importantly, it is extremely resistant to mold. Other recommendations include the "shimmering, sometimes skunky" Raspberry Parfait, which is available as an autoflower (see Photoperiod Flower vs. Autoflower section below), and the "unusual and floral terpenes" of Stargazer, which is resistant to Septoria.

Seeds vs. Clones

Cannabis can be propagated vegetatively by rooting cuttings. Both this method, cloning, and propagation from seed are viable at any scale, which is very unusual for a crop species. There are many different options for sourcing seeds and plants and just as many different opinions, and in the end, it is a personal choice based on what is available to you and your skill level. But for a home-scale grower who is only growing a few plants, Ben recommends starting with high-quality feminized seed from a reputable seed producer, such as Humboldt Seed Company. The advantage here is that you will be almost guaranteed to have a female plant (which produces the flowers) instead of a male plant (which does not). There is a way to grow un-feminized seed and then rogue out the male plants when they start showing male characteristics, but this takes a bit of skill. Since most home growers only need two or three plants, it is usually worth spending the extra money on good feminized seed. Buying clones of known varieties is also an option, but plants started from seed tend to be healthier and more vigorous. For advanced growers, cloning your own plants to keep the genetic line going is a free way to produce more plants but requires year-round effort and a controlled environment.

Photoperiod Flower vs. Autoflower

All of the original drug-type cannabis genetics were photoperiod dependent, which means the plants would only begin to flower when the amount of sunlight per day decreased to a certain threshold. For these varieties, flowering, ripening, and harvest will always occur around the same time in the fall in a given location. Recently, however, breeders have developed a plethora of autoflower varieties, which flower after a set number of days regardless of sunlight conditions. There are advantages and disadvantages to both, but regardless, Ben highly recommends staggering the flowering window of your plants. For example, if you are growing three plants, you may want to choose varieties that will be ready to harvest in September, October, and November. This will increase your chance of getting a dry harvest window and also spread out the work of post-harvest drying and processing, which is the most time-consuming part of growing cannabis. If you add autos to your garden, you can grow any autoflowering strain and plant in succession to add July, August, and maybe even June to your harvest calendar in temperate zones!

Harvest and Uses

Proper harvest and curing of cannabis depends on what the end use goal is and, like growing, is an entire art and science unto itself. However, the basic principles are to harvest the flower at the peak of ripeness and then dry away from direct sunlight. For flowers, the buds go through an additional curing process after drying for best results.

As you probably know, there are almost too many historical and potential uses of the cannabis plant to list! While this chapter is focused mainly on smokeable flower, we are also excited about the so-called industrial uses of hemp such as building materials and textiles. In addition, there is an emerging (and also ancient) world of medicinal cannabis use. While some of the medicinal varieties are THC-dominant strains like the ones mentioned here, there is also an array of alternative cannabinoid and ratioed strains being developed, including the very popular 1:1 THC to CBD strains. These strains are worth exploring particularly if you are interested in the non-psychoactive medicinal effects of the plant. There is also a welcome move towards extracting and using the cannabis plant so that it can be consumed without smoking. Tincture extractions are a great way to enjoy the medicine, as are cannabis leaf teas (non-psychoactive). Ben and Betsy are especially excited about their infused honey made by combining honey from their farm with a solventless extract from their cannabis flowers. Yum!

4. STRAWBERRY

Trevor's Story

Wild strawberry was the first plant I learned to fear. In downtown Richmond where I grew up, plants were for beauty and aesthetics, or else they were the mysterious wild weeds that sprang up in sidewalk cracks and at the edges of abandoned urban lots. The backyard next to mine was one such lot—a neglected patch of dusty dirt, weeds, and "junk" trees—but it was my favorite place to play. As young as five or six, I was drawn there almost subconsciously. When no one was watching me, I'd hop over the rickety wooden fence separating our manicured backyard from the neighbors, and I'd lose myself in the magical privacy of that world, searching for sticks and rocks and "roly-poly" bugs. But most of all, I remember, I liked the way it *felt* in there. I liked the peace of being in a forgotten place that everyone else had abandoned. I liked the mystery that hung like magic dust in the air, the mystery of unknown plants and beings, of an emerging wildness that seemed to have a life of its own. I liked the way the sunlight danced through the jumble of leaves and branches and onto my skin. I liked that there were indescribable things in that space that it seemed like only I had ever discovered. And I loved the little thumbnail-sized red strawberries that carpeted the ground in June. My parents had told me forcefully that they were poisonous and not to eat them, which added a powerful tinge of fear to the awe I already felt. Every summer, I looked forward excitedly to finding those strawberries and incorporating them into my magical world.

Years later, soon after I had moved west and started learning about native plants and foraging, I was on a plant walk with a mentor, and we came across a patch of tiny strawberries growing on the edge of the trail. He began to tell me about the huge genus *Fragaria* that includes dozens of strawberries, and then he declared that all of the strawberries worldwide are edible. He bent down and picked a strawberry, ate it, and then handed one to me. Even though I knew intellectually it was okay to eat it, I still felt hesitant. I ate two of the berries. They weren't very delicious, and as my mentor told me, some of the wild strawberries can be almost bland while some can have intense flavor. We kept hiking, and it wasn't until an hour had passed and I didn't feel sick at all that the mild fear of that wild strawberry finally went away for good.

I don't fault my parents for telling me the strawberries were poisonous. Whether they thought they were or not, it's a sensible strategy to tell a young child not to eat anything in the wild, especially in the city where most plants with shiny colorful fruits probably are inedible. But for me, it was part of a larger disconnection from nature that stemmed from growing up in that peculiar modern American urban/suburban culture of the 1990s. It was a mindset and a cultural attitude, and it pervaded everything from the landscape to mass media to my public school. It was why I didn't grow a vegetable until I was 20. It was why I didn't learn the name of a tree until I was 18. And it's why almost everything I've done in my life as an adult has one way or another been geared towards reconnecting to nature, both for myself and for my children.

A few years after my foraging mentor taught me about wild strawberries, I was camping in the Appalachian woods with a friend. We were doing an all-day ramble, not intentionally fasting, but being more focused on immersing ourselves in the forest, we hadn't packed any food. We were taking a sit break in the middle of a thick patch of woods when my friend popped up from the forest floor and announced with excitement, "There's wild strawberries everywhere!" I got down on the ground with him and sure enough, underneath the thick cover of ferns and sprouted tree seedlings and moss was a carpet of tiny wild strawberries, stretching for hundreds of feet in all directions. And while these were tiny, they were packing a lot of flavor! We spent the next couple of hours crawling around, munching on tiny delicious strawberries, stopping to lay on the ground and look at the treetops occasionally, feeling for a bit like we belonged to the forest and it belonged to us.

For the next several years, too busy to forage or go to the local strawberry patch to pick, I was forced to buy strawberries from the grocery store. They were so bland and mealy that I stopped buying them altogether. Even the top-dollar organic strawberries bought in-season were unpalatable and had an aura of "I was grown with exploited labor and giant machines in dusty, dead 'certified organic' California fields."

With the magic of the strawberry sadly fading in my life, I decided to take matters into my own hands and grow a big patch on our new farm. It was actually one of the first crops I put in that fall after we moved onto the land (strawberries are usually planted in September), and we were so busy that winter taking care of our infant that I had almost forgotten about them come springtime. The deer had done a lot of damage to the plants over the winter, and I was worried we wouldn't get much yield, but when it warmed up in April, the plants took off, and by May 1st they were in full flower, and by June 1st, we were swimming in a sea of luscious red strawberries.

And oh were they sweet! I know it's annoying to brag about your own produce, but no joke, a single one of these strawberries packed the same amount of flavor as three quarts of Driscoll's. And what's more, not a single thing, organic or not, had ever been sprayed on them. It felt good to know that the heavenly fruits we ate daily around the breakfast table were made up exclusively of soil, sunlight, and rain.

And there was one more delightful surprise to that wonderful 2019 strawberry crop. My son, Afton, was ten months old when the strawberries were ripening up, and he was refusing to eat real food. My wife and I tried strategy after strategy to no avail, and were getting to the point of looking into next steps, feeding specialists, etc. Then one day in mid-May, we were all down in the strawberry patch together, and we turned around to see Afton sitting on the ground with a giant strawberry in his mouth, red juice smeared across his face, grinning blissfully. Parents who have experienced the heightened stress and worry of first-time child-rearing can imagine the elation and relief we felt as we watched while he finished the entire strawberry (green sepals and all) and enthusiastically hand signaled for more! From that day on, Afton has been obsessed with berries of all kinds, and most especially the ones we let him pick straight from the farm.

How to Grow

The strawberry family (*Fragaria* spp.) contains a wide variety of perennial low-growing species from all over the world, including several that are native to North America. The cultivated, or garden strawberry, is usually a hybridized species that is bred to predictably bear large fruit in full-sun conditions. Our native North American strawberry, *Fragaria virginiana*, by contrast, grows in the forest understory and produces tiny (but delicious) berries.

To grow a good crop of garden strawberries, you will need relatively rich, well-drained soil in full sun. Although strawberries are a perennial, we like to grow ours as an annual crop, which is the way they are typically grown commercially. We've adapted the commercial system into a planting cycle for our homestead to get the high yields of the annual system and the efficiency of the perennial system.

Our planting cycle starts in the first week of September when we plant our healthy strawberry plugs (2" minimum size) into a prepped, bare 30" wide garden bed. We plant a staggered double row down the length of the bed at 14" spacing within the row. When the temperatures start to go below 20 degrees in the winter, we put some protective agricultural fabric over the row, which we remove in March. The plants will begin to flower in the spring and set fruit in late April or early May. To get good yields, we aim to have the strawberry plants grow to the size of a dinner plate before the winter sets in the first year after planting, which in our climate means we have to plant by September 15th.

In the late summer, we snip off the strawberry runners and root these "daughter" plants in plug trays in the greenhouse or under shade cloth. These plugs will be what we use to plant out the next bed of strawberries come September.

Meanwhile, we use the original bed of strawberries to interplant a summer crop like tomatoes or hemp or a fall crop like cabbage—something large that can be transplanted directly into the bed of strawberries.

Integrating into Your Plant Guilds

Now, I know what you're thinking—aren't strawberries a perennial?! The answer is yes, of course, and you should definitely experiment with growing a perennial patch of strawberries. We like the annual method because we get predictable large harvests and also have plenty of space to continually plant more patches. To keep a perennial patch fruiting requires a bit of pruning, rooting of new daughters, and constant weeding. However, on a small scale or around the base of an island guild tree planting, it is definitely worth growing them as perennials.

Cultivar Recommendations

In our climate and soil, the two best cultivars are Chandler and the everbearing Albion. If you want to grow a small patch of strawberries that won't spread, such as in a container planting, Alpine is a good choice. Other standout varieties include Earliglow and Galleta from North Carolina State University.

5. HONEYBERRIES

Trevor's Story

Sometimes I think I became a fruit grower because of how hard it was to find good-quality, flavorful berries growing up in the mass-produced food era of the 1990s. I've loved berries since I can remember—as a kid who wasn't allowed to eat much candy, they were the closest thing, and healthy—but they were always a hit-or-miss affair in my household. Was this batch of strawberries that my Mom got from the store going to be yummy, or taste like cardboard? Are these blueberries going to actually have that satisfying pop and zing of a fresh berry, or will they collapse into bland mush in my mouth?

On the rare occasion that the store-bought berries in our fridge were tasty, I had to contend with yet another obstacle to my dream of gorging on delicious berries. My Dad had a thing where he liked to put berries on his Cheerios in the morning, and he had a rule that we were only allowed to eat the berries as an add-on to cereal, just like him. One day I found a colander full of primo blueberries sitting on the kitchen counter, and looking around to make sure I was alone, I began to devour them by the handful. Oh the bliss, this was a really good batch! But then, right as I was reaching into the colander for another helping of berries, my Dad arrived in the room and snatched them away, roaring, "Onlyyyy for the cerealllll!!"

Now, I get it. My Dad was raising three ravenous kids, store-bought berries are really expensive, and a 12–year-old can eat $20 worth of berries in no time. He had a point. But right then and there, the seed of my berry farming dreams was planted—how amazing it would be to have unlimited access to fresh, no-spray, nutrient-dense berries at the peak of ripeness.

Fast-forward to my early farming days. I was disappointed to find out that many of the berries I loved were really hard to grow organically in our climate, especially the beloved blueberries. (It turns out the berries I ate growing up had probably been soaked with all kinds of pesticides and fungicides.) In my quest for easy-to-grow, no-spray berries, I began to hear about the honeyberry, which was unfamiliar to me. The size and taste of a blueberry crossed with a grape, high in the same antioxidants as blueberries, pest and disease resistant. Sounds amazing! The only problem was that no one I knew was growing them in my region, and some said it was too warm to grow them.

Determined, I was fortunate to be awarded a grant from the Claypool-Lehman Research Fund at the North American Fruit Explorers (NAFEX) to trial ten different honeyberry varieties on my farm. They took a few years to start producing, and we had to figure out how to get them to fully ripen, but last year we cracked the code on growing them and were blessed with a banner year of honeyberries. That May and June, we would go out to the orchard every evening and eat fresh honeyberries right off the bush, and I finally got to tell my kids, "Go ahead, eat as many as you want, we have plenty."

How to Grow

The honeyberry (*Lonicera caerulea*), also known as haskap, just might be the most underrated berry to grow for those of us in colder temperate climates (Zone 7 and colder). They ripen extremely early (even before strawberries), are pest and disease resistant, and most importantly are delicious fresh out of hand, with the flavor and texture of a blueberry crossed with a Concord grape.

Honeyberries have traditionally been grown and bred in very cold climates such as Russia and northern Japan, so there is probably a limit to how warm of a climate they can tolerate. We have had a lot of success growing them in Zone 6, even after many people told us it was too warm to grow them in our area. One reason we have been successful is by trialing new varieties to see what works here. In addition, we have planted ours on a north-facing slope in the partial shade of a Food Forest guild, and we irrigate them on very hot days in the summer to keep their roots cool. If you are in Zone 6 or warmer, we recommend using these techniques to grow this cold-climate crop. If you are in Zone 5 or colder, you should be fine growing them in full sun similarly to blueberries.

Integrating into Your Plant Guilds

In terms of placement in a growing system, honeyberries are extremely versatile due to their small size (four feet tall by four feet wide when mature), ability to grow in partial shade, and wide soil adaptability. As such, they are the perfect shrub to integrate between trees in a linear guild Food Forest planting. With semi-dwarf trees spaced approximately 20 feet apart in a row, you can easily fit two or three honeyberry shrubs in between each tree. We have not tried it, but we have heard of people top grafting honeyberry onto Japanese honeysuckle bushes, as they are in the same family. If you have wild honeysuckle around, you might consider trying this experiment and reporting back!

Cultivar Recommendations

We have been conducting a cultivar study here at Wild Rose Orchard over the past few years, funded graciously by the North American Fruit Explorers (NAFEX), and soon to be published in the NAFEX publication *Pomona*. Overall, we have found that all of the varieties we planted grew well and fruited on our farm, but there is an overall winner in terms of flavor—the variety Aurora, bred by the University of Saskatchewan. Aurora was the clear favorite in many blind taste tests on our farm, so much so that we recommend planting at least 50% of this cultivar for fresh eating. However, be sure to include other cultivars in your planting for cross-pollination. Other favorites from our trials include Boreal Beauty, Boreal Beast, Tana, and Keiko.

Please note there are dozens of honeyberry varieties out there with a range of ripening times, sizes, and flavors. Different cultivars may very well grow better in your climate. For example, some excellent work has been done by the late Maxine Thompson at Oregon State University to breed cultivars specifically for the Pacific Northwest. One great resource for growing information as well as plant stock is HoneyberryUSA.com, an online nursery in Minnesota.

Harvest and Uses

One common mistake to warn you about—if you pick your honeyberries too early, they will be sour and you may decide you don't like them! I believe this is one of the reasons the berries have not caught on faster in areas where they have not traditionally been grown. Leave the berries on the bush as long as you can and at least until the juice is dark purple. When truly ripe, they are very sweet and tangy and have a soft velvety texture! We have not had a problem with birds in our honeyberry patch, although many growers report that they are forced to put netting around their berries in order to allow them to completely ripen before the birds eat them.

Honeyberries are excellent for fresh eating straight off the bush and are therefore a hit with children and a great candidate for a U-Pick crop similar to blueberries. But they also have endless applications for value-added and frozen delights, including jams, jellies, syrups, smoothies, and as a topping on your favorite breakfast dish or dessert. Amazingly, honeyberries have the highest flavonoid content of any of the most commonly eaten berries—twice the amount as blueberries! The majority of that antioxidant power comes in the form of anthocyanin, the compound found in blue-pigmented fruits and vegetables that is particularly good for protecting against cardiovascular disease.[1]

CHAPTER 7

Dynamic Accumulators

1. COMFREY

Ryan's Story

SOMETIMES, on late nights after too much wine, a group of friends and I will find ourselves sitting around the fire inevitably bashing as many Permaculture stereotypes as we can. Herb spirals, rain barrels, and swales all get a proper roast. We've all had our herb spiral era, haven't we? Despite or because of our own successes and failures, it is an easy thing to lay into these predictable tropes. Comfrey is at the top of this list.

Perhaps because it is so easy. Comfrey was in fact the first plant I bought for my own Permaculture site. This was 20 years ago, and now my site is loaded with thousands of comfrey plants. It's so easy to grow and spread that I feel bashful selling it to anyone. My response to most people when they ask about comfrey is, "Come dig it up."

Here's the thing—it is so multifunctional. For years I did the chop and drop thing, convinced that I was depositing and spreading the dynamically accumulated minerals from deep in the earth onto the surface. In my guilds, I made sure to include the tough plant right next to many fruit trees. There has since been some concern about a carcinogen that makes comfrey more suspect with regard to this function. At SPI we are paying attention to the research as it develops.

Another use I loved was as a supplemental feed for the chickens. They love comfrey! For years, I hung out with this plant, slowly using it when necessary, but this relationship remained surface. Its stature in the landscape seemed to conjure in me a "meh." Meanwhile I'd read about the plant, and it would shine. I just couldn't get into it.

Right around the same time of my swale, rain barrel, and herb spiral era, I literally slipped into a much deeper relationship with comfrey.

I live in a small 900-square-foot log cabin. Back in the early 2010s, I was juggling running a nonprofit farm, being the full-time caregiver for our newborn son, getting my own vegetable farm off the ground, and picking up shifts pouring wine on weekends and shifts as a clinical mental health counselor after hours in order to make the nonprofit paycheck math make sense. For most of my day, my son followed me everywhere in a BabyBjörn. It was common for me to be standing up rocking him with a bottle in his mouth during a board meeting. Point is, it was a hustle. I had very little time in the day, and I could not afford to take off any time.

One Tuesday morning, I was halfway through my first morning cup of coffee, checking emails from one of my jobs, when I heard my son, Tamayo, letting out his first morning cry. It was time for me to spring into action and start the morning routine—diapers, bottles, snuggles. My log cabin has wooden stairs that on the one hand are charming, on the other are a nightmare in socked feet. I had just snapped the onesie shut on my son and picked him up to rush downstairs to finish up the morning before we had to head off to a meeting with a potential donor.

My first step onto the steps in a hurry proved to be the one. My socked foot slipped, and in the time it takes to snap your fingers, both of my feet were "ass over teakettle" as they say.

My baby son was still in my arms. I had a decision to make. How was I going to land?

My son was in my right arm cradled like a football. I needed to fall in a way that avoided crushing him with my body weight. I rolled my body to the left just enough to take the full hit with my left elbow. All 250 pounds of me down a full staircase.

It hurt. My son just giggled.

We moved through the rest of the morning. Everything felt normal except for the goddamned throbbing pain that kept shooting through my left arm. I didn't think anything of it. As a former college basketball player, the mantra "Are you hurt or injured?" constantly played in the back of my head, and I was convinced this was just pain.

The problem popped up when we got to the part of my day where it was time to farm. I took off my long sleeve to get ready for the hours of hand weeding ahead, and that's when I noticed my elbow had blown up like a balloon. The next realization—I couldn't straighten or bend my left arm.

I couldn't farm.

Fortunately, that afternoon I had a friend coming by who happened to be an herbalist and acupuncturist. She was visiting in order to pick up starts from my greenhouse for her annual garden. When she opened the door to the greenhouse, she quickly noticed my one arm working with a wince on my face.

"What did you do to yourself?"

I told her the story.

"Let me have a look."

When she took my elbow into her hands, I immediately noticed a light bulb go off in her eyes.

"Is it broken?" I asked. "Is my season over?"

"You burst your bursa sack, follow me."

She led me from the greenhouse back to my house where she had me sit on my couch. On the way, she reached down and ripped out a large double handful of spring-growth comfrey. In the house, she put the pot on and started cooking it down into a poultice. I was skeptical but appreciated this friend taking the time to care for me.

After cooking down the comfrey and letting it cool for a bit, she brought the entire stewpot of comfrey sludge over to the couch and directed me to soak my elbow in it.

The relief was remarkable. The feeling of soaking the injured arm in the poultice had a satisfying feeling of release, as if the pressure I previously felt was quickly leaving the body. There was pain, but it was as if the poultice was sucking the pain out of my elbow.

She left shortly after with instructions for me to soak my arm several more times that evening. The next morning when I awoke, my elbow had returned to its normal size and I had full mobility. I could farm!

This event is the moment I finally understood what comfrey has to offer. The plant showed up for me in a big way thanks to a friend and a burst bursa sack.

How to Grow

Comfrey (*Symphytum* spp.) is a low-growing herbaceous perennial from Eurasia that grows in full sun and partial shade in a variety of soil types from Zone 3 to 9. While it is a tenacious plant that can get established just about anywhere, it will be more productive in rich, fertile soil that stays well watered, and in hot climates it may benefit from growing in partial shade. In addition to being a dynamic accumulator, it also produces beautiful purple flowers that are a favorite nectar source for bees.

Integrating into Your Plant Guilds

Comfrey is a dynamic accumulator of potassium and silicon and also a prolific grower of biomass, making it a must-have plant for the herbaceous layer of any guild or orchard.[1] Comfrey spreads out vegetatively over time, so you do not need to plant a lot of it. We recommend one plant around every few anchor trees to get started. It grows fast, producing large soft leaves that grow in clumps. To keep it from taking over too quickly and to speed up the soil-building process in your guild, chop and drop the plant throughout the season by cutting the leaves at the base and spreading them around as natural mulch. Typically we chop and drop our dynamic accumulators twice throughout the growing season.

Variety Recommendations

We recommend planting Bocking 14 or Russian Comfrey (*Symphytum uplandicum*), which is a hybrid cross that produces sterile seeds and will keep comfrey from spreading beyond where you plant it. Common comfrey (*Symphytum officinale*) is considered invasive because the seeds will spread far and wide, and once it gets established somewhere, it is very hard to eradicate.

Propagation

Of all the plants in our orchard, comfrey is the easiest to propagate from root cuttings. A single piece of the root as small as an inch, pushed into the ground, is enough to start a new plant. Comfrey's ability to root and grow quickly is what makes it such a powerful plant for the orchard. You can start with one clump and propagate it from root cuttings out to all of your guilds. However, just like sunchokes and nettles, comfrey is very hard to remove once established. This is definitely a plant for the larger confines of the orchard rather than the kitchen garden.

Harvest and Uses

Comfrey, although not edible, is an important component of the herbal first aid kit for external skin and wound care. Comfrey leaves contain a compound called allantoin that is found in many natural skincare products because of its ability to stimulate the growth of new skin cells. For this reason, it is one of our favorite ingredients in homemade healing salves and skin lotions along with other common garden herbs such as calendula and borage. A poultice or wrap of the leaves can be used to heal bruises and sprains, but do not use on open cuts. Comfrey is such a powerful healer that it could cause wounds to close up too soon and trap infection!

2. BEET

> *The beet was Rasputin's favorite vegetable.*
> *You could see it in his eyes.*
>
> —Tom Robbins

Ryan's Story

Here and Now

I have very recently become enamored with the color pink. All things pink. My collection has begun: a pink scarf, pink hat, pink cell phone. Pink seems to permeate and fill every crevice, every moment. Before heading further down this pink path, I think first it necessary to communicate to you, dear reader, a simple fact that may make my obsession with pink even more interesting. I am in the midst of dusting the cobwebs off the old hypothalamus and venturing into some of the loftier areas of the pituitary glands. Yes, oiling up that crazy wisdom once again and taking a spin around the axons and dendrites of the soul. This road out of a depressive episode can feel like a May morning after popping a few sunrise Adderalls. Strange new energy, a little exercise, and often a greatest hits of nostalgic wins logged during moments from the before times.

Picasso had his blue period and so did I, except that only I'll remember it because I'm not Picasso and besides the blue really had a little more of a green tint to it, as did Picasso, if you count the absinthe-soaked flashes he indulged in, since we're on the subject of colors. The peculiar thing is that, somewhere along my seratonic climb, I hopped and skipped the brightening of the Matisse blue, landing me right onto a plateau of soft Barbie pink.

Given my newly discovered appreciation and, yes, affection for all things pink, you can understand, I'm sure, the shock and excitement I experienced when I sat down to a bowl of borscht, of all things, and suffered a crippling flashback so powerful I might as well have been perched atop a lookout shooting at very real-though-imagined planes after watching some poorly done sensationalized Vietnam-era movie on bad acid.

None of this was intentional, mind you. The borscht, a treat brought to my house by a friend, at first struck me as pink, for sure, but more of a bright cranberry pink. The neon pink of the red-light district establishments that no longer feel the need to flash the red lights that once signaled their membership in this lusty club.

It is the pink of danger. This all changed of course when I spooned a dollop of sour cream into the bowl, and the once demure come-hither of a pink gave way to a soft swirl of a candy cane embrace, and it happened. I flashed and backed into a world not of Southeast Asian police actions or bad acid but one riddled with cancer and Rasputin.

My mother had a Hungarian friend. Before Mom got sick, she spoke of her often. She escaped something from somewhere and fled to Charlottesville for some reason. The word was that people were looking for her, and we had to keep her country of origin a secret. That was the word, but Mom never kept secrets well.

Turns out, the runaway Hungarian was actually Russian, and misassumed to be Anastasia—*the* Anastasia, rumored to be hiding out in Charlottesville all these years. Those damned Blue Ridge Mountains always made me nervous. It would be nice to know it was just the ghost of Rasputin hovering over the Blue Ridge barking out pink coughs. I am tempted to claim this. But that would be wrong because this tale is about borscht.

Our Hungarian friend, probably Russian and probably not Anastasia, made us borscht once. Fucking borscht. For a year or two, I watched lasagna after lasagna land on our stovetop in a parade of well-meaning meal trains. Mom was fighting cancer, and the lasagna got so boring—cold, sitting on the stovetop—that we all opted for cold hot dogs instead. Only the Tupperware changed. Except for the borscht. The borscht seemed to stay singular in a bowl in our kitchen forever and never changed colors. This pink, to my irrational eight-year-old mind, saved my Mom. (Of course it didn't, but don't tell 1985 Ryan that.)

The embedded virtues in pink vegetables are not a subtle one. Just ask Tom Robbins. Yellow of the onion, red of the pepper or purple of tomatoes, and white-green of the celery stalk are great, all things being equal. If the lot were to share a stewpot, you would eventually find the celery stalk, through a wilting head of hair no doubt, whimpering in the corner with her rear end painted not red but pink of all colors. This, the throbbing pink of the here and now.

Something Rasputin understood.

This leaves us with the centerpiece of borscht. The beet. It is my favorite to grow. And, in my dreams, it cures cancer.

How to Grow

The humble but mighty beet is way more versatile than it gets credit for. It can be grown from early spring until late fall in all temperate climates and has multiple uses as a culinary delight, staple crop, fodder crop, and cover crop. While it does prefer light loamy soil, it will grow anywhere along the sand-to-clay soil spectrum. We like to start our succession of fresh-eating beets in the spring when we plant our first round of lettuce, greens, and carrots. Then we sow a succession of beets every two to three weeks to make sure we always have some to harvest. In the late summer, we plant one large succession of storage beets that we will harvest when the weather turns cold and store them in the root cellar for the winter.

Mangel beets, or sugar beets, can be planted in a fodder block along with other high-calorie fodder crops for animal feed, especially for pigs and chickens. We have found that mangels are also a great addition to a cover crop mix for the way they open up and de-compact the soil.

Integrating into Your Plant Guilds

While most of our favorite guild plants are perennials, we sometimes like to plant annuals in our guilds, especially in the first few seasons as our trees are getting established. If you have already tilled or sheet-mulched an area of soil to make room for planting your guilds, you might as well use that opportunity to get an annual yield, since preparing new ground is so labor-intensive. For example, let's say you have a six-foot by six-foot circle-shaped island guild that you sheet-mulched the previous year. You have your Asian pear tree ready to plant in the center, but you haven't propagated any shrubs or herbaceous perennials for the rest of the guild. Why not plant annual vegetables in the rest of the circle to keep the soil covered and energized with living roots and grab a quick yield? We like to plant either spreading plants like sweet potato and winter squash, or root crops like carrots and beets.

Variety Recommendations

We like all beets, but for fresh eating, especially on salads, we love the golden beets like Touchstone Gold and the red-and-white striped varieties such as Chioggia.

Propagation

Beets are best started by seed although some market gardeners do start them in trays and transplant them. One funny quirk about beets is that each seed usually contains two true seeds, so you will often have to thin out the extra seedlings that germinate. Saving seed from beets is easy and worthwhile. Choose one open-pollinated variety to save seed from and let it grow through the first season, and in the second season, it will flower and set seed. As long as you only let one variety go to seed, you can save the seeds and they will be mostly true to type.

Harvests and Uses

Beets are an underrated storage crop, able to be stored for six months or more in the right conditions. There are many different methods, but we like to store ours covered in sand in a cool moist environment like a root cellar (this is also the same way we store carrots). You will want to grow a storage variety such as Lutz Winter Keeper and time it towards the end of the summer so you can harvest in the cooler conditions of the early fall. Another way to store beets in cold climates is to sow them as late as possible to get a good yield and just leave them in the ground when winter sets in. This way, you can just go out and dig some up any time throughout the winter when the ground isn't frozen and they will be perfectly fine.

Borscht Soup

Ingredients:

4–7 beets, peeled and grated
4 tablespoons olive oil, divided
8 cups chicken broth
3 pinto gold fingerling potatoes, sliced into small pieces
2 carrots, sliced into small pieces

Broth base:

2 celery ribs, trimmed and finely chopped
1 onion, finely chopped
1 large heirloom tomato

Additional flavorings:

2 bay leaves
2–3 tablespoons white vinegar
1 teaspoon sea salt, or to taste
¼ teaspoon black pepper, freshly ground
1 bulb of hardneck garlic (we prefer the Appalachian heirloom variety called Music)
3 tablespoons chopped fresh dill

Directions:

Peel, grate, and/or slice all vegetables (keeping sliced potatoes in cold water to prevent browning until ready to use then drain).

Heat a large soup pot (5 ½ quart or larger) over medium/high heat and add 2 tablespoons olive oil. Add grated beets and sauté 10 minutes, stirring occasionally until beets are softened.

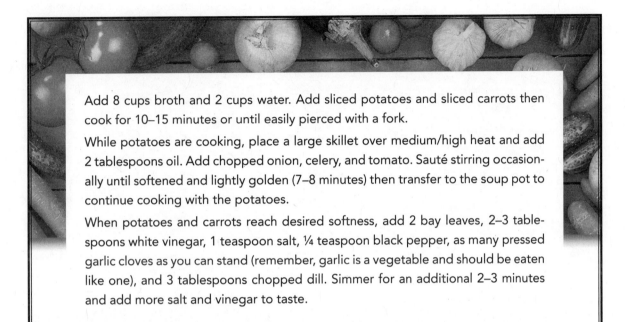

Add 8 cups broth and 2 cups water. Add sliced potatoes and sliced carrots then cook for 10–15 minutes or until easily pierced with a fork.

While potatoes are cooking, place a large skillet over medium/high heat and add 2 tablespoons oil. Add chopped onion, celery, and tomato. Sauté stirring occasionally until softened and lightly golden (7–8 minutes) then transfer to the soup pot to continue cooking with the potatoes.

When potatoes and carrots reach desired softness, add 2 bay leaves, 2–3 tablespoons white vinegar, 1 teaspoon salt, ¼ teaspoon black pepper, as many pressed garlic cloves as you can stand (remember, garlic is a vegetable and should be eaten like one), and 3 tablespoons chopped dill. Simmer for an additional 2–3 minutes and add more salt and vinegar to taste.

3. NETTLES

Trevor's Story

Everyone's got a fungophile in their lives, that person completely and utterly obsessed with fungi who has the wild shroomy personality to match. Mine is Charlie, aka "Charliceps." We cut our teeth learning to forage wild fungi together while living up on the Appalachian Plateau, but very quickly he surpassed me in his drive and knowledge and became my fungal mentor. He taught me how to take spore prints, how to interpret the litany of idiosyncratic mushroom ID words in our guidebooks (fibrillose, ovoid, viscid when wet), and how to anticipate where to hunt based on weather, topography, and forest composition. Gradually he helped me gain enough confidence to eat the edible mushrooms I found, and we were soon bringing in hauls of delicious chanterelles, boletes, and lactarius to sauté in butter and feast on with our friends. I came to trust his expertise so much that I was even willing to eat the mysterious Amanita Jacksonii mushroom with him—a delicious edible that most people stay away from because it is in the otherwise poisonous Amanita genus (think death caps and destroying angels).

Charlie also has that eccentric personality that many a fungophile seems to have, which made our epic hunts all the more fun. On our rambles, he never wore shoes, of course, and for some reason, he was always leaving his underwear in random places in the woods and finding them a few days later (I can't explain it...!). To this day when I picture him, I see him climbing a tree in the woods, barefoot and shirtless, with a mesh sack of mushrooms on his back (to seed the forests with spores) and honey locust thorns in his earlobes. We used to cruise around looking for "mushies" in his old Volvo station wagon that was always filled with everything you would ever need for a spontaneous adventure plus a lot of things you wouldn't—buckets of acorns, blacksmithing gear, old containers of Split Banana gelato, and lots and lots of mushrooms drying on the dashboard.

Eventually we moved to different areas, and our paths diverged. I moved onto a farm and had a baby. He stayed in the city and started a mushroom business. I missed him, of course, and I also started to feel like we had less and less in common. Mainly this was because I was getting rusty on my mushroom ID and was feeling a little self-conscious about it. So, when he invited me on a chanterelle foray one July, I immediately said yes, but as the day grew closer, I began to feel a little nervous.

My nervousness disappeared when he pulled a classic Charlie move at the outset of the trip. He knew I had a newborn in the house, so when he arrived, he quietly knocked on the door so as not to wake the potentially sleeping baby. My son was at his grandparents that day, and I was actually inside catching up on sleep myself, waiting for Charlie to arrive and wake me up. I didn't hear him when he knocked and didn't answer. Charlie took this as a sign that the baby was asleep, so he just sat there on the front porch and didn't knock again and waited … for three hours! No joke. He eventually knocked on the door when he heard rustling inside and meekly told me he had been outside since nine and it was now noon. I asked him what he had been doing the whole time, and he said with complete seriousness, "Practicing my Karate katas." It was hilarious and also touching. We both got a huge laugh out of this all the way up the road to the woods.

When we slid off the trail and into the forest at Braley Pond, Charlie immediately started finding mushrooms. This was always how it went, so I was used to it. But as time passed and we meandered through valley after valley, I began to get frustrated that I wasn't spotting anything. Charlie's sack was full of beautiful golden chanterelles, and mine was empty save a couple of oyster mushrooms I had found growing on a fallen log. Mushroom hunting isn't a competition, but it is a collaborative effort and I felt like I wasn't pulling my weight.

After a while, we came to a fork in the valley, and without discussing it, we split up. Charlie went left and I went right. This is one of the things I love about our forays together—they are improvised rambles without a plan or timeline, and we often split off on our own adventures and eventually reconnect somewhere else in the forest, hours later. That day I found a small, shaded ravine and followed my curiosity upstream into the holler. As the earth narrowed, it grew more lush with vegetation, and I found myself standing in a huge patch of wood nettles—knee-high plants with clusters of pale yellow flowers popping out of their tops like confetti. I had just learned about this native relative of stinging nettles in my herbalism class. I knew it was a nutritive powerhouse just like stinging nettles, but without stingers, it was a breeze to harvest a bunch to cook down as edible greens. I went to work filling my sack with the younger nettles that had not yet flowered, which I knew would be the most tender and delicious.

Charlie and I wandered into each other back towards Braley Pond about an hour later and excitedly compared our hauls. With renewed pride, I showed him the beautiful wood nettles I had collected. He didn't know the plant so I told him all about it and suggested we sauté it up with vinegar and eat it for dinner along with our mushrooms. He was stoked to learn a new plant and have another wild foraged dish for our feast, and I was just happy that I was able to contribute something. Here the guild metaphor is irresistible—a cliché but true. I realized that just like in a well-functioning guild, everyone has something different but unique to contribute to the system. I love foraging mushrooms, but I will never be the expert Charlie is. For some reason, I have always been called to plants. They speak to me and drive me to obsession, and obsession develops over time into expertise. Specialization can be a really fun and healthy thing as long as it is embedded in a community of competencies.

How to Grow

Stinging nettle (*Urtica dioica*) is an herbaceous perennial native to Eurasia that has naturalized across much of the world and has a long history of edible and medicinal use. In recent years, it has also been studied for its benefits as a dynamic accumulator, particularly of calcium. All of this, coupled with the fact that it is incredibly easy to grow, make it a top plant for any guild.

Nettle can grow in a range of different soils and in full sun or partial shade. However, it does seem to struggle in the heat of the summer. In our south-facing full-sun garden, it is the first herb to emerge in the spring, putting on lots of vibrant growth through April. When it gets hot in May or June, it will shrivel up and stop growing until the fall brings cooler temperatures, at which point it will flush with growth again.

It is called *stinging* nettle because it is covered in small hairs or spines that can be very irritating to the touch (unless you're Ryan, who seems to get a masochistic thrill out of rubbing stinging nettles over his body). The irritant is actually a chemical that will subside after a few minutes and that gets broken down and becomes inert when cooked.

Integrating into Your Plant Guilds

Nettle spreads vegetatively faster than any plant we have in our garden, which can be a blessing and a curse, depending on where you put it. We recommend growing a nettle patch away from frequently used areas and in a place where you can mow the perimeter to keep it in check. When planted in a guild, it might take over, but that's okay by us, you just probably don't want to plant it in every guild.

Nettle is a superstar dynamic accumulator because of how fast it grows and cycles nutrients; you can chop and drop it multiple times throughout the season. It seems to be particularly adept at moving calcium from the subsoil to the topsoil, and one study showed that liquid fertilizers made from nettles contained relatively high concentrations of seven key nutrients, including potassium and magnesium. What's fascinating is that preparations of nettle have long been used in many different agricultural traditions to enhance plant health, including biodynamics. We like to include nettles along with several other plant extracts in some of our holistic sprays in the orchard.

Variety Recommendations

We're not aware of any improved varieties of nettle—like many of our favorite Permaculture plants, they are so easy to propagate that we recommend snagging cuttings from a friend. It is interesting to note, however, that wood nettle, the nettle in Trevor's story, is in the same family as stinging nettle and is native to eastern North America. *Laportea canadensis* grows in the shady understory of deciduous forests and looks very similar to nettle, but the hairs on its stems do not sting. We haven't tried it, but perhaps wood nettle could be grown in the full shade of a mature Food Forest as an analog to its stinging cousin.

Propagation

Stinging nettle can be grown from seed or transplants but is also incredibly easy to propagate from cuttings. It spreads so rapidly that you can dig out sections of a patch for propagation and they will fill back in in no time.

Harvest and Uses

The key to using nettle for culinary purposes is to cook it first. The heating process will break down the stinging chemicals and render them harmless. It is a great addition as a cooked green along with classics like kale and collard or other wild cooked greens like dandelion. We like to combine it with garlic mustard greens and cook them down with vinegar—the sweet flavor of nettles balances the pungency of the garlic mustard.

To enjoy the medicinal properties of nettle, we add it to herbal tea decoctions of all types. Our favorite is a nutritive spring tonic that includes nettle, mountain mint, anise hyssop, and motherwort.

If we are harvesting nettles to eat or for tea, we like to pick them when they are young before the plants flower. My favorite method, since I don't like wearing gloves but also don't like to get stung too much, is to pinch the top six inches off each shoot, where there are less stinging hairs than lower on the plant.

Later in the summer, the nettles will flower and go to seed and are no longer as good to eat, but this is the time to harvest them to use in your biological sprays in the orchard.

4. BURDOCK

Ryan's Story

We were young, my wife Joy and I, working multiple jobs, and putting everything we had into finding a piece of property in the area that we grew up in so that we could, in the words of Wes Jackson, "find a place to become native to." A four-acre rocky knoll, surrounded by conventional corn, wheat, and soy on all sides, is where we landed. Fueled by Toby Hemenway's book, *Gaia's Garden*, I was determined to design and develop the piece of property into our permanent family Utopia.

We moved onto the property that spring and got to work getting to know the place. Those early years were hard work, and we had zero idea what we were doing. Gardens failed, grass grew into thistle patches out of our well-meaning intention of allowing "nature" to just do its thing.

Despite reading a few books, the idea of partnering with succession had yet to sink in. Instead, I had stopped fighting succession and allowed it to take over. This resulted in a plant I had not yet identified taking over the property.

This plant loved to grow giant leaves that shot up little purple flowers turning into little balls of sticky seeds that would grab hold of anything that brushed past it and be carried to its next place. When we went outside, every time we got back in the house, our clothes were full of burrs.

One night, things turned scary. We were sitting at the kitchen table talking, enjoying our Saturday night, when my daughter, who was a crawling age, grabbed onto my leg and looked up at me with fear in her eyes.

She was choking. I panicked and lifted her up to the table. My wife and I helped her try to swallow by patting her back and turning her on her chest. Soon she was turning blue. That's when complete terror set in.

In a desperate move, my wife jammed her finger down my daughter's throat and pushed the blockage through. My daughter recovered from the fear, and we finished the night enjoying each other and processing the event.

We also wondered what it was that she had choked on. Soon after, in the diaper, clear as day, was one of those damned burrs. Fear turned to anger, and I committed the very next year to eradicate the plant from my property.

Enter Dr. Ted Butchart

Ted was a naturopathic doctor in town who also happened to be a straw bale builder and a Permaculture designer/educator.

When we first met, he agreed to come tour the property and chat about our shared love of plants. We became fast friends, and it was Ted, during this walk on my property, who helped change my mind about burdock.

We were walking the edge of my property, and I was complaining about the conventional agriculture operation next door. I was profoundly worried about the herbicide drift and the health impact this could have on my family. I shared that the spraying happened twice a year.

In an effort to help me relax, he was explaining how this wasn't the end of the world when he stopped short, pointed, and said, "There, that's your solution!"

He was pointing at burdock. I was confused. To me and my wife, this was a junk plant we wanted to get rid of. Dr. Ted immediately turned poetic and walked me through a rather mystic wondering of how incredible it is that land seems to grow just the plant that can be a solution for the "poison" that is infecting it.

He went on to explain the medicinal benefits of the plant and how it, along with dandelion and cleavers—also prevalent on the property—is a liver and blood cleanser. That day, we dug up some roots, and he taught me how to make a tincture out of it in his apothecary. From then on, I was enchanted by burdock and set out to learn and get to know this plant as closely as possible.

How to Grow

Burdock (*Arctium lappa*) is an herbaceous biennial native to Eurasia that is naturalized across temperate North America. In our region, it's a common weed that almost everyone would recognize for its large elephant ear leaves and Velcro-like burrs that stick to clothes. Unbeknownst to many people nowadays, it is a powerful medicinal plant as well as an edible root vegetable with a long tradition of herbal and culinary use around the world.

Burdock grows wild throughout our farms, especially recently disturbed areas. Every season, we leave patches of burdock plants unmowed and mark them with flags so we will be able to find them for harvest in the late fall.

If you don't have wild burdock where you are—or if you just want to grow an abundance of the crop, especially as an edible vegetable—you can grow it in your garden along with your other root crops. The key to growing large burdock roots and harvesting them effectively is to plant in loose, fluffy, de-compacted soil. The best way to de-compact garden soil is to use a heavy-duty broadfork over several seasons, in combination with lots of compost and good organic growing practices.

Some people grow burdock in raised containers to make it easier to harvest. In Japan, where it is a popular culinary vegetable, it is grown in special boxes with removable sides to make harvest easy. (By the way, this same method can be used on a small scale for easy production and extraction of sweet potatoes and potatoes.)

Burdock is best grown by direct seeding the seeds into the soil where they will grow, rather than transplanting from the nursery. This will allow the taproot to get established in the beginning and result in the largest and longest roots.

Integrating into Your Plant Guilds

We consider burdock a dynamic accumulator in our guilds simply because its taproot is so extensive and deep and its leaves so large and fast-growing. Like comfrey and rhubarb, the leaves can be chop and dropped several times throughout the season with little negative effect on the underlying root crop. We also allow burdock to grow naturally in wild strips throughout and next to our orchard. The idea is to designate strips of ground as no-mow zones and let your native and naturalized perennials grow up. These become both mini-habitats for beneficial insects and easy areas to forage a plethora of edible and medicinal plants like burdock, dandelion, dock, plantain, etc.

Variety Recommendations

If you are growing burdock in the garden, instead of wild harvesting, we recommend purchasing seeds of improved cultivars, most of which were bred in Japan. Baker Creek Heirloom Seeds has a variety called *Takinogawa* that was first developed in Japan over 300 years ago and grows roots over three feet long.

Harvest and Uses

Burdock root has a sweet, earthy flavor and is a great addition to stir-fried dishes, especially when paired with carrots. For traditional Japanese recipes, search the Japanese name of the vegetable, which is called *Gobo*.

We use wild foraged burdock root for its tremendous medicinal properties. As Ryan learned from our friend Dr. Butchart, the root is a great addition to spring cleansing tonics along with dandelion root and milk thistle.

Each year in the late fall, we locate the burdock plants that we flagged during the summer and methodically dig them out to collect the roots. Take your time and try to get the entirety of the long taproot. It is best to harvest in the fall, or winter after the first year of growth because this is when the most nutrition is stored in the roots. To do this, flag plants in the summer that do not have flowers or seeds yet, then come back and harvest them in the fall or winter. (Burdock is a biennial that grows only leaves in its first year and flowers and goes to seed in its second year.)

After washing and drying the burdock root, we dice some of it into small pieces to make a burdock root tincture. We slice the rest into thin strips and air-dry in the sun then store away for use in teas. Both the tincture and tea can be used year-round to cleanse the system, but it is especially beneficial in the spring as we are all coming out of our own hibernation and starting to become more active.

The way we always remember when and how to use burdock is by referring to it as Bear Medicine. Like the bear coming out of a sluggish winter, it helps get the liver working again, breaks down fats in the digestive system, and generally cleanses the lymphatic system and purifies the blood. (Sure enough, the Latin name of the genus *Arctium* is derived from the root word for bear!)

5. YARROW

Trevor's Story

Every plant lover has that first plant that just stumps them again and again. For me that was yarrow. My yarrow trials started when I moved to the mountain farm in Virginia during the early days of my plant education. The fields there were filled with an abundance of useful plants like goldenrod, dandelion, plantain, as well as two plants that I could not for the life of me tell the difference between—yarrow and Queen Anne's lace. Both had small feathery leaves, both had tiny white flowers clustered together in an umbel, but each had very different uses. How was I supposed to tell the difference?!

The breakthrough for me came when I learned that Queen Anne's lace is in the carrot family (in fact it is the wild ancestor of the cultivated carrot), which gave me the idea to pull up the root. Indeed it looked, smelled, and tasted like a carrot. Then I took some leaves and crushed them up in my hand and smelled them—carroty indeed! Yarrow, on the other hand, smelled totally different, very medicinal and distinct. I had discovered my identification key, the miraculous and often underappreciated olfactory sense. From that day on, smell became an important tool in my plant and mushroom ID tool kit (it is especially useful with mushrooms that often have very distinct odors between similar-looking species).

Nowadays, I can recognize yarrow or Queen Anne's lace from the corner of my eye, like noticing an old friend passing by on the street. It's hard to even imagine how I could mistake the two, given yarrow's much finer and more intricate leaves and completely different growth form. So much of plant ID is about pattern recognition, and learning patterns comes down to exposure and repetition. I have to remind myself that there was a time that the patterns of nature were like a foreign language to me, just like the patterns of music still are. I was late to the journey myself, having not started until my early 20s, which makes me wonder, how much deeper is there to go? What patterns will emerge in the decades to come? What depth of knowledge might I have if I'd been raised embedded in nature like my ancestors?

In 2015, I had the opportunity of a lifetime to travel to my ancestral homeland in Lithuania with my Mom and some other family members. My great-grandparents on my Mom's side were born in Lithuania and lived on a farm in a small village outside of the city of Kaunas. It was a life-changing experience, meeting distant relatives who I had never met before (they all looked just like my Mom!), foraging for mushrooms in the ancient forests of my fore-elders, and learning about Lithuanian culture and the deep reverence they hold for nature. It suddenly became so clear where my love of nature came from. It was embedded in me like it is in all of us, passed on through the DNA and cultural memory of my ancestors. And to actually be *in the place* where the symbiotic relationship evolved was so powerful. I felt a profound sense of belonging that I have yet to feel anywhere else on planet Earth.

Towards the end of our trip, my mom and I were walking along a river, the banks lush and overgrown with wild plants. There on the banks, I noticed my new (old?) friend yarrow, a continent and ocean away from the mountain where we first met. "Mom!" I said, "this is yarrow, it grows back home too!" And so it does.

How to Grow

Common yarrow (*Achillea millefolium*) is a highly variable species that defies the simplistic dichotomy of "native" vs. "non-native." It most likely originated in Eurasia but long ago spread throughout North America and adapted to almost every environmental niche, from the alpine slopes of the Rockies to the coastal plain of the Mid-Atlantic. What all of this means for us growers is that yarrow has a place in almost every single farm and garden in the temperate world and is incredibly easy to grow and propagate!

Integrating into Your Plant Guilds

Yarrow is a sun-loving herbaceous perennial that grows two to three feet tall in the summer and produces tiny white flowers that are a favorite insectary for beneficial insects like parasitic wasps. It is also a must-have for most guilds and Food Forest plantings because of its multifunctional uses.

In addition to being a beneficial attractor in your guild, it is also an effective barrier plant to prevent grass from encroaching. To achieve this, plant several yarrow seedlings side by side in a tight clump around the outer edge of the guild. Make this protective moat of yarrow at least a couple of feet deep around the guild. An aggressive grower, it can be cut back mid-season after flower stalks have finished blooming to encourage more herbaceous growth and even a second bloom cycle. As the yarrow grows over multiple seasons, it will spread vegetatively and self-seed, maintaining a nice barrier for several years.

Propagation

Yarrow is easy to start from seed and can also be propagated by root division, like rhubarb or comfrey. The best way to get new yarrow plants is to get a large clump established and just dig out sections of it whenever you need them. It will quickly spread or self-seed to fill in the gap you made. If you get ahold of yarrow seeds they are very easy to start, germinating readily along with your other flower and vegetable seeds in the nursery.

Variety Recommendations

The best way to get yarrow started in your garden is to find a healthy patch growing wild in your area and propagate that, or grab some from a friend or neighbor. This way you will be starting with genetics that are already partially adapted to your area. There are even ornamental varieties of yarrow available with different colored flowers including pink, red, and yellow, although we are not certain that these are as resilient as the more wild varieties.

Harvest and Uses

Yarrow is a premier medicinal plant for human and animal uses, and a whole book could be written about the historical and modern herbal uses of the amazing plant (hint: check out the mythological story behind the Latin name!). For us, yarrow has become nature's Band-Aid, and we have planted it all over the place so it is always at hand as part of our living first aid kit. When one of the kids gets a cut, they now know to pluck a yarrow leaf and fold it into the shape of a Band-Aid and then wrap it around the cut. The amazing thing is that it has medicinal properties that both stop the bleeding *and* disinfect the wound! We also like to dry the leaves and flowers for use in herbal teas throughout the season. Our favorite blends are an elderflower and yarrow tea for colds and flus and a yarrow, red clover, burdock, and dandelion blend for spring cleansing.

We Close With a Metaphor

"Trevor, you're being an asshole," Ryan said.

"I don't care," Trevor retorted.

Emilie just shook her head.

We were stuck. Emilie, Trevor, Ryan.

There we were, seated outside a fancy coffee shop in Charlottesville, Virginia. On the lawn, under a tree. It was 96 degrees in the shade, and the three of us were arguing again. Across the street in view sat a hedge.

"What is that plant?" Ryan asked, quickly pausing the argument, distracted by curiosity and the shared love of plants.

"Witch hazel." Emilie speaks up, eager to cut the tension.

"Huh? I don't know that plant," Ryan says.

"Oh man, it's great, the plant blooms in the dead of winter like a giant middle finger to everything else that is dormant," Trevor continued. "And the best part, it's high in tannins—incredibly bitter. It's the tannins that make a healing salve."

The three of us look at each other, the spell of the disagreement having been broken,

"Ryan's bitter ass needs to plant a whole bunch of those on his property."

Emilie couldn't resist the opportunity to roast Ryan.

The three of us look at each other and break out into laughter there on that lawn. As passersby wonder, what sort of strange inside joke could have been that funny.

Farming, writing a book together, owning a permaculture business, these things are all a lot like marriage. It's radical belonging. Intimate. We know each other at our best and at our worst.

It's a lush feeling, abundant, even safe when you can love a person, love a community, when things are tough.

Upon reflection, this book may feel like a gritty read. All it really is are stories about life. As the stories of people and plants intersect, it becomes apparent—the plants, if we let them, help us to make meaning out of our human experience.

And therein lies the pearl.

When we started writing this book, we had a pretty loose idea of what it was about. We knew how important plants were in our own lives, and we wanted to share that, along with a few things we've learned along the way. What we learned as we wrote was that the plants aren't just supporting curiosities, they are central characters. They're as important to our stories as the people. They're another layer of the web that supports us, lifts us up in joy, and catches us when we fall.

The significance of that witch hazel for us wasn't that it saved us or solved our conflict, it was just the reminder that it was there. As metaphor, as food, as medicine—and that realization alone is enough to shift things. They make it all make a little more sense. As Gregory Bateson says: "It's the difference that makes a difference."

Now go find *your* witch hazel, or rhubarb, or mulberry. Then report back. You're part of our community now.

Endnotes

Introduction
1. Carol Nash, "Native American Communities of the Shenandoah Valley: Constructing a Complex History," 2020, https://bpb-us-e1.wpmucdn.com/sites.lib.jmu.edu/dist/9/133/files/2019/04/Native-American-Communities-of-the-Shenandoah-Valley.pdf

Chapter 1: The Human Sector
1. Jack H. Presbury et al., *Ideas and Tools for Brief Counseling*, Merrill Prentice Hall, 2002.

Chapter 2: Food Forests Simplified
1. James Shackell, "Could These Ancient Food Forests Help Feed the World?" July 15, 2019. https://adventure.com/ancient-food-forests.
2. Sebastian Salinas-Roco et al., "N2 Fixation, N Transfer, and Land Equivalent Ratio (LER) in Grain Legume–Wheat Intercropping: Impact of N Supply and Plant Density," *Plants,* 13, no. 7, March 2024.
3. Ben Tyler, "New Findings Further the Study of Dynamic Accumulators," Cornell Small Farms Program, April 2022, smallfarms.cornell.edu/2022/04/new-findings.
4. Fedco Seeds, "Michael Phillips' Holistic Spray Recipe," https://smallfarms.cornell.edu/2022/04. Accessed July 31, 2024.

Chapter 3: Anchor Plants
1. Eliza Greenman, "Mulberry Preamble," 2019, elizapples.com/2019/10/31/mulberry-preamble.

2. Trees for Graziers, "Persimmon: Seven Months of Manna from Heaven." treesforgraziers.com/persimmon-seven-months-of-manna-from-heaven. Accessed July 31, 2024.
3. Memorial Sloan Kettering Cancer Center, "American Pawpaw: Purported Benefits, Side Effects & More," 2021, www.mskcc.org/cancer-care/integrative-medicine/herbs/american-pawpaw.

Chapter 4: Nitrogen-Fixers

1. Béla Keresztesi, "Breeding and Cultivation of Black Locust, *Robinia Pseudoacacia*, in Hungary," *Forest Ecology and Management*, 6, no. 3 August 1983, https://doi.org/10.1016/S0378-1127(83)80004-8.

Chapter 5: Barrier Plants

1. Richard A. Kralj, "Rhubarb: Love It for Its Taste, Eat It for Your Health," Penn State Extension, 2023, extension.psu.edu/rhubarb-love-it-for-its-taste-eat-it-for-your-health.
2. Patsy Evans, "Reducing the Japanese Beetle Population," UConn Today, 2015, today.uconn.edu/2015/06/reducing-the-japanese-beetle-population/.

Chapter 6: Beneficial Attractors

1. H.P.V. Rupasinghe et al., "Short Communication: Haskap (*Lonicera caerulea*): A New Berry Crop with High Antioxidant Capacity," *Canadian Journal of Plant Science,* 92, no. 7, November 2012, cdnsciencepub.com/doi/10.4141/cjps2012-073.

Chapter 7: Dynamic Accumulators

1. Tyler, Ben. 2022. "New Findings Further the Study of Dynamic Accumulators—Cornell Small Farms." Cornell Small Farms. https://smallfarms.cornell.edu/2022/04/new-findings-further-the-study-of-dynamic-accumulators/

Index

Bold page numbers indicate plant growing information, *italic* page numbers indicate diagrams.

A

access, in SPI Scale of Permanence, 22
aesthetics, in SPI Scale of Permanence, 23
American persimmon (*Diospyros virginiana*), **83–85**
  hunting and, 80–83
anchor plants
  about, 46–47
  American persimmon, 83–85
  Che tree, 78–79
  mulberry, 69–70
  pawpaw, 88–90
  willow, 72–74
annuals in guilds, 174–175
Anthropocene, 15
Asian persimmon, 85
asparagus, propagation, 118
autumn olive (*Elaeagnus umbellata*), 111

B

barrier plants
  about, 47
  elderberry, 137–139
  garlic, 127–129
  Jerusalem artichoke, 122–123
  recommendations, 62–63
  rhubarb, 117–119
  spring bulbs, 132–134
Bateson, Gregory, 192
bee sting story, 107–108
beet, **174–175**
  borscht, 172–174, 176–177
beneficial attractors
  about, 48
  cannabis, 153–156
  honeyberry, 163–165
  recommendations, 62–63
  strawberry, 159–161
  tulsi, 144–145
  zinnia, 148–149
black locust (*Robinia pseudoacacia*), **104–106**
  as firewood, 103–104
  fritters, 98
Borscht Soup, 176–177
bread, pawpaw, 90
Brown, Brené, 146–147
buildings, in SPI Scale of Permanence, 22–23
burdock (*Arctium lappa*), **185–187**
  medicinal use, 183–184
Butchart, Ted, 5, 184
Byrd, Susanna, 148–149

C

cannabis (*Cannabis* spp.), **153–156**
  legalization of, 150–152
capital, 8 forms of, 28–35, *29*
The Challenge of Landscape (Yeomans), 19
Che tree (*Maclura tricuspidata*), **78–79**
  connection to, 75–78
cheat sheet, for Food Forests, 61–63
cherry silverberry. *See* goumi
chicken in-and-out exercise, 35–36
Chinese melonberry. *See* Che tree
circulation, in SPI Scale of Permanence, 22
Clarke, Jenna, 33
climate, in SPI Scale of Permanence, 20
clover (*Trifolium* spp.), **108–109**
  bees in, 107–108
comfrey (*Symphytum* spp.), **170–171**
  guild categories of, 48
  medicinal use, 167–170
  propagation, 118
common yarrow. *See* yarrow
community, of plants, 42, 192
COVID, effect on Permaculture education, 13–14
Crawford, Martin, 56–57
Creating a Forest Garden (Crawford), 56
cultural capital, 31

D

daffodils. *See* spring bulbs
Dancing Star Farm, 73, 120–122
Dancing Star Farm Sunchoke Chips, 124
daylilies. *See* spring bulbs
Daylily Wine Festival, 130
design goals, 45
  See also patterns
disease management, 64–65
diversity, in Permaculture education, 13–14
Dusen, Adam, 88–89
dynamic accumulators
  about, 48–49
  beet, 174–175
  burdock, 185–187
  comfrey, 170–171
  nettle, 180–182
  recommendations, 63
  yarrow, 189–191

E

Ecology of Self model, 23–26
ecosophy, 24, 25
ectomycorrhizal fungi, 74
Edible Landscaping, 79
8 Forms of Capital, 28–35, *29*
elderberry (*Sambucus* spp.), **137–139**
  as connector, 135–137
  Emilie's Elderberry Syrup, 140
expanding island guilds, 57–58, *58*
experience of place, in SPI Scale of Permanence, 23
experiential capital, 31

F

false indigo. *See* indigo bush
fermentation, Lacto-Fermented Garlic, 129
Ferver, Buzz, 84
financial capital, 28–30
fishing story, 91–96
Food Forests
  development framework, 42–43
  implementation of, 49–52
  layers, *44*
  perception of, 39–42
  plant recommendations, 60, 62–63

site selection and layout, 52–53
using guild framework, 43–49
versatility of, 3
foraging, 4–5, 157–159, 178–180
forest gardening. *See* Food Forests
fritters, redbud, 98
fruit trees. *See* trees
fruiting shrubs, recommendations, 62
functions, stacking, 36
fungi foraging, 178–180

G
garlic (*Allium sativum*), **127–129**
　Lacto-Fermented Garlic, 129
　as transition, 125–127
garlic chives, 128
goumi (*Elaeagnus multiflora*), **111–113**
　as chicken habitat, 110–111
　Goumi Jam, 113
Grocery Aisle concept, 59–60
groundhog, Whistle-pig au Vin, 32, 34
guilds. *See* plant guilds

H
haskap. *See* honeyberry
Head, Heart, Hands exercise, 36–37
herbaceous layer, *41*, 63
The Holistic Orchard (Phillips), 64
holistic sprays, 64–65
holistic thinking, 25
holy basil. *See* tulsi
honeyberry (*Lonicera caerulea*), **163–165**
　trialing, 162–163
Hori Hori, 135–137
human in-and-out exercise, 35–37
Human Sector
　definition, 16
　Ecology of Self model, 23–26
　effect on Permaculture projects, 18–19

8 Forms of Capital, 28–35, *29*
framework of, 3
guild metaphor in, 46
Head, Heart, Hands exercise, 36–37
human in-and-out exercise, 35–36
LUV triangle, 26–28
in Scale of Permanence, 20–21
in zones, 16–17
humans
　discovering relationship with plants, 14–15
　impact on land, 23–24
　relationship with plants, 3

I
Indigenous peoples, 8–10, 41–42
indigo bush (*Amorpha fruticosa*), **101–102**
in-home counselor story, 99–101
infrastructure, in SPI Scale of Permanence, 22–23
in-home counselor story, 99–101
inner landscape of human energy, 23–26
input-and-output exercise, 35–36
intellectual capital, 30
interpersonal observations, 24–25
intrapersonal observations, 24
irises. *See* spring bulbs
island guilds, 57–58, *58*

J
Jacke, Dave, 17–20, 45
jam, goumi, 113
Jerusalem artichoke (*Helianthus tuberosus*), **122–123**
　Dancing Star Farm Sunchoke Chips, 124
　digestive issues with, 120–122
Jones, Mark, 73
Judd, Michael, 57

K
Keillor, Garrison, 115

keyline pattern, 55
Kraft, Chuck, 107–108
L
Lacto-Fermented Garlic, 129
landform, in SPI Scale of
  Permanence, 21
Landua, Gregory, 28
Lawton, Geoff, 41
layers, in guilds, 44–45
layout, of site, 52–53
linear guilds
  layout of, 53–55, *54*
  spacing, 55–56, *56*
listen, 26–27
Lithuania, 189
living capital, 30
Long Acre Farm, 138
LUV (Listen Understand Validate)
  triangle, 26–28
M
Mamma Shirley, 107–108
mangel beets, 174
material capital, 31
McConkey, Michael, 79
McKee, Ed, 26
medicinal plants
  burdock, 184, 186–187
  cannabis, 156
  comfrey, 169–170, 171
  rhubarb, 119
  tulsi, 145
  yarrow, 190–191
mental health professionals,
  99–101, 146–148
Mentore, Laura, 15
metaphors, 27–28, 191–192
microclimate, in SPI Scale of
  Permanence, 22
mint, 45–46
Miracle Farm, 55
Mollison, Bill, 25, 32
mulberry (*Morus rubra, M. alba,
  M. nigra*), **69–70**

introduction to, 67–69
mushroom foraging, 178–180
N
Naess, Arne, 25
Nepal, 40–41
nettle. *See* stinging nettle; wood
  nettle
nitrogen-fixers
  about, 47
  black locust, 104–106
  clover, 108–109
  goumi, 111–113
  indigo bush, 101–102
  recommendations, 62
  redbud, 96–98
North American Fruit Explorers
  (NAFEX), 163, 164
nurseries, 66
O
O'Neill, Dave and Lee, 54, 116
orchards, definition, 43
  See also Food Forests
Oriental strawberry. *See* Che tree
P
"Painted Leaves," 80–83
Parfitt, Tom, 97
patterns
  approaches to, 59–60
  for expanding island guilds,
    57–58, *58*
  for linear guilds, 54–55, *54*
  mimicking nature, 43
  plant interactions and, 3, 4
  simple, 44–45
pawpaw (*Asimina triloba*), **88–90**
  foraging, 4–5
  memories of, 86–87
  Trevor's Pawpaw Bread, 90
peonies. *See* spring bulbs
Perfect Circle Farm, 84
Perkins, Richard, 55
Permaculture
  definition, 8, 10–11

human element of, 5
perception of, 39–40
Permaculture Design Courses (PDCs)
  authors' experience with, 16–19
  diversity in, 13–14
  effect on relationships, 130–132
Permaculture educators, 6
Permaculture plant, definition, 58–59
pest management, 64–65
Phillips, Michael, 64, 65
plant guilds
  about, 3, 43
  expanding island guilds, 57–58, *58*
  Food Forest implementation, 52–53
  functional categories, 4, 45–49
  linear guilds, 53–56, *54, 56*
  patterns, 44
  plant recommendations, 62–63
  plant selection, 59–60
plantain, 109
plants
  community of, 42, 192
  criteria for Permaculture plant, 58–59
  history with Indigenous ancestors, 9–10
  recommendations, 62–63
  relationships with, 3, 11, 14–15, 42
  selection of, 59–60
  sources of, 66
  value of, 1–2
popcorn disease, 70
potlucks, 32–33
Project GROWS, 5–6, 7, 33, 35

Q
Queen Anne's lace, 188

R
Radical Roots Farm, 54, 116
redbud (*Cercis canadensis*), **96–98**
  fishing story, 91–96
  Redbud Fritters, 98
regenerative agriculture, 8
relationships
  discovery of, 14–15
  in Food Forest cultivation, 42
  observations of, 24–25
  Permaculture and, 11, 130–132
  value of, 191–192
rhubarb (*Rheum × hybridum*), **117–119**
  versatility of, 115–116
Ridgedale Permaculture, 55
rootstocks, 62
Rowland, Ethan, 28, 30

S
Samuelson, Ben, 153–156
Sand Talk (Yunkaporta), 10
Scale of Permanence
  author introduction to, 17–19
  evolution of, 19–20
  human-centered, 20–23
Seed and Soil Farm, 153
seed catalog ritual, 14
seed companies, 66
self-seeding plants, 145
self-talk, 24
Shenandoah Permaculture Institute
  development of, 5–7
  human-centered Scale of Permanence, 20–23
Shenandoah Valley, 9–10
shrubs, *44*, 62
site selection and layout, 52–53
Sobkowiak, Stefan, 55, 59–60
social, in SPI Scale of Permanence, 20–21
social capital, 28–30, 35
social Permaculture, 16–17, 18–19
soil fertility and management, in SPI Scale of Permanence, 23

Soul Valley Wizard, 103–104
soup, borscht, 176–177
spiritual capital, 31
spirituality, 25
spring bulbs, **132–134**
  impact of, 130–132
Spring Creek Blooms, 148
stakes, 64
Staton, Renee, 26
stinging nettle (*Urtica dioica*), **180–182**
  foraging, 178–180
strawberry (*Fragaria* spp.), **159–161**
  foraging, 157–159
sugar beets, 174
sunchoke. *See* Jerusalem artichoke
Sunchoke Chips, 124
syrup, elderberry, 140

T
Taylor-Jones, Jenny, 135–137
Tiphia wasps, 133
Tonesmeier, Eric, 116
Traditional Ecological Knowledge, 8–10
transpersonal observations, 24, 25
trees
  as anchor plants, 46–47
  canopy layer, *44*
  preparation and planting, 64
  recommendations, 61
  rootstock, 62
  spacing, 55–57, *56*
Trevor's Pawpaw Bread, 90
Trice, Betsy, 32–33
tulips. *See* spring bulbs
tulsi (*Ocimum tenuiflorum*), **144–145**
  as adaptogen, 141–144

Tweardy, Emilie
  development of SPI, 5, 6
  human-centered Scale of Permanence, 20–23
  potlucks, 32
  relationships, 191–192

U
understand, 27–28

V
valerian root, 120–121
validate, 28
vegetation, in SPI Scale of Permanence, 22
Virginia, Indigenous peoples, 9–10

W
water, in SPI Scale of Permanence, 21–22
water remediation, 73
Whistle-pig au Vin, 32, 34
Whitefield, Patrick, 11
Wild Rose Orchard, 46, 49–52, 53
wildlife, in SPI Scale of Permanence, 22
willow (*Salix* spp.), **72–74**
  memories of, 71–72
Wilmer, Cody and Paige, 138–139
witch hazel, 191
wood nettle (*Laportea canadensis*), 180, 181–182

Y
yarrow (*Achillea millefolium*), **189–191**
  identifying, 188–189
Yeomans, P.A., 17, 19, 55
Yunkaporta, Tyson, 10

Z
zinnia (*Zinnia* spp.), **148–149**
  therapy and, 146–148
zones, for designer, 16–17

About the Authors

RYAN BLOSSER is a farmer, educator, writer, mental health professional, co-founder of Shenandoah Permaculture Institute, and owner/operator of Dancing Star Farm. The intersection of his passion for growing food and helping people fuels his unique perspective on building community resilience through permaculture design. Ryan lives in Churchville, Virginia.

TREVOR PIERSOL is a farmer, permaculture designer, educator, co-founder of the Shenandoah Permaculture Institute, and owner/operator of Wild Rose Orchard. Trevor is dedicated to the advancement of regenerative agriculture with a particular focus on perennial fruits, medicinal herbs, and easy-care native plants. He lives in Mount Sidney, Virginia.

ABOUT NEW SOCIETY PUBLISHERS

New Society Publishers is an activist, solutions-oriented publisher focused on publishing books to build a more just and sustainable future. Our books offer tips, tools, and insights from leading experts in a wide range of areas.

We're proud to hold to the highest environmental and social standards of any publisher in North America. When you buy New Society books, you are part of the solution!

- This book is printed on **100% post-consumer recycled paper,** processed chlorine-free, with low-VOC vegetable-based inks (since 2002)
- Our corporate structure is an innovative employee shareholder agreement, so we're one-third employee-owned (since 2015)
- We've created a Statement of Ethics (2021). The intent of this Statement is to act as a framework to guide our actions and facilitate feedback for continuous improvement of our work
- We're carbon-neutral (since 2006)
- We're certified as a B Corporation (since 2016)
- We're Signatories to the UN's Sustainable Development Goals (SDG) Publishers Compact (2020–2030, the Decade of Action)

At New Society Publishers, we care deeply about *what* we publish—but also about *how* we do business.

To download our full catalog, sign up for our quarterly newsletter, and learn more about New Society Publishers, please visit newsociety.com.

ENVIRONMENTAL BENEFITS STATEMENT

New Society Publishers saved the following resources by printing the pages of this book on chlorine free paper made with 100% post-consumer waste.

TREES	WATER	ENERGY	SOLID WASTE	GREENHOUSE GASES
34 FULLY GROWN	2,700 GALLONS	14 MILLION BTUs	110 POUNDS	14,500 POUNDS

Environmental impact estimates were made using the Environmental Paper Network Paper Calculator 4.0. For more information visit www.papercalculator.org